WITHDRAWN FROM STOCK

2 0 FEB 1987

COLLECTING AND RESTORING
SCIENTIFIC INSTRUMENTS

Other books by Ronald Pearsall
The Worm in the Bud
The Table-Rappers
Victorian Sheet Music Covers
Collecting Mechanical Antiques

COLLECTING AND RESTORING SCIENTIFIC INSTRUMENTS

Ronald Pearsall

DAVID & CHARLES
NEWTON ABBOT LONDON

0 7153 6354 9

© Ronald Pearsall 1974

All rights reserved. No part of this publication may be reproduced, stored in a retrieval system, or transmitted, in any form or by any means, electronic, mechanical, photocopying, recording or otherwise, without the prior permission of David & Charles (Holdings) Limited

Set in 11/13pt Baskerville
by Trade Linotype Limited Birmingham
and printed in Great Britain
by Biddles Ltd Guildford
for David & Charles (Holdings) Limited
South Devon House Newton Abbot Devon

CONTENTS

LIST OF ILLUSTRATIONS   8
ACKNOWLEDGEMENTS   12
INTRODUCTION   13
1 SURVEYING INSTRUMENTS   20
    Groma — Astrolabe — Waywiser — Odometer — Pedometer — Measuring chain — Measuring rod — Theodolite — Circumferentor — Cross staff — Optical square — Vernier scale — Graphometer — Quadrant — Octant — Common theodolite — Plane tabling — Sector — Altazimuth theodolite — Levels — Division engines — Repeating circle — Compasses
2 NAVIGATIONAL INSTRUMENTS   50
    Compass — Cross staff — Mariner's astrolabe — Nocturnal — Octant — Sextant — Artificial horizon — Chronometer — Station pointer — Hydrometer — distance-finder
3 DIALS   67
    Temporary hours — Equinoctial hours — Diptych dial — Butterfield dial — Ring dial — Crescent dial — Equinoctial dial — Spherical dial
4 TELESCOPES   74
    Refracting telescope — Chromatic aberration — Compound eyepiece — Reflecting telescope — Types of mirror — Achromatic lens — Army telescope — Clockwork drive — Silvered-glass mirror

CONTENTS

5 SPHERES AND ORRERIES 100
Celestial globe — Gore — Armillary sphere — Terrestrial globe — Orrery — Pocket globe

6 SPECTROSCOPES 109
Spectroscopy — Spectroscope — Spectroscope with camera — Use with Bunsen burner — Small direct-vision spectroscope — Micro-spectroscope

7 MICROSCOPES 116
Simple microscope — Compass microscope — Screw-barrel microscope — Lieberkühn — Compound microscope — Use of adjusting screw — Cuff microscope — Achromatic objective — Binocular microscope — Solar microscope — Reflecting microscope — Lever stage movement

8 BAROMETERS 134
Torricellian experiment — Cistern barometer — Siphon barometer — Wheel barometer — Two-liquid barometer — Diagonal barometer — Folded barometer — Conical barometer — Three- and four-liquid barometer — Baroscope — Portable barometer — Marine barometer — Fortin barometer — Aneroid barometer

9 THERMOMETERS 146
Galileo's thermoscope — Liquid-in-glass thermometer — Clinical thermometer — Fluorescent thermometer — Mercury-in-steel thermometer — Adjustable-range thermometer — Gas- and vapour-pressure thermometers — bimetallic thermometer — Electrical-resistance thermometer — Byström's pyrometer — Hot-blast pyrometer

10 PRECISION BALANCES 154
Suspended balance — Harrison balance — Ramsden balance — Pneumatic balance — Spring balance

11 DRAWING INSTRUMENTS 159
T-square — Set square — Parallel ruler — Protractor — Compasses — Beam compasses — Bow compasses — Dividers — Pantograph

CONTENTS

12 SURGICAL INSTRUMENTS 164
Scalpel — Surgical saw — Needle-holder — Forceps — Retractors — Scissors — Probes — Ophthalmoscope — Laryngoscope — Lithotrite — Tonsillotome — Ophthalmometer — Inhaler — Electrolysis — Cautery — Cystoscope — Urethroscope

13 MATERIALS AND MANUFACTURE OF SCIENTIFIC INSTRUMENTS 171
Ivory — Gold and silver — Wood — Copper printing plates — Brass — Metals — Silver plating — Electroplating — Instrument makers — Machine tools — Lathes — Screws — Turret lathe — Accurate scales

14 CLEANING AND RENOVATION OF SCIENTIFIC INSTRUMENTS 182
Cleaning brass and copper — Lacquering — Gilding — Renovation of other metals — Cleaning wood — Treating veneers — Removal and application of veneers — Ageing of new wood — Coverings — Mirrors — Lenses — Mechanical repairs

GLOSSARY OF SCIENTIFIC INSTRUMENTS 197

NOTES ON PRICES 238

EIGHTEENTH-CENTURY INSTRUMENT MAKERS 245

SCIENTIFIC-INSTRUMENT MAKERS OF LONDON IN 1843 AND 1894 253

BIBLIOGRAPHY 268

INDEX 272

LIST OF ILLUSTRATIONS

Trade cards and advertisements (*Science Museum*) 14
Simple quadrant (*Science Museum*) 15
Theodolite (*Science Museum*) 16
George Adams of Fleet Street, London (*Science Museum*) 17
Early Victorian trade card (*Science Museum*) 18
Astrolabe (*Science Museum*) 21
Mid-eighteenth-century waywiser (*Sotheby & Co*) 23
Negretti and Zambra instrument (*Phillips*) 23
Pattern of triangulation and surveying compass and a protractor (*British Cyclopaedia* 1835) 26
Ramsden theodolite (*Science Museum*) 27
Circumferentor (*Science Museum*) 29
A sturdy cross staff (*Science Museum*) 30
Mid-eighteenth-century brass surveying instrument (*Sotheby & Co*) 30
Vernier scale (*Dictionary of British Scientific Instruments* 1921) 31
Graphometer 32
Simple quadrant (*Science Museum*) 33
Graphometer of Italian design (*Science Museum*) 34
Sturdy kind of compass (*Science Museum*) 35
Plane-tabling resulting in first-rate maps (*Smith Elder & Co*) 36
A variety of sectors (*Science Museum*) 37
Early nineteenth-century theodolite and two levels (*British Cyclopaedia* 1835) 39

LIST OF ILLUSTRATIONS

A level incorporating a telescope (*Science Museum*) 40
Ramsden's dividing engine (*British Cyclopaedia* 1835) 41
Complicated gearing and machinery (*British Cyclopaedia* 1835) 42
Troughton and Simms theodolite (*Science Museum*) 42
Dollond's reflecting circle and Troughton's reflecting circle (*British Cyclopaedia* 1835) 44
Troughton reflecting circle (*Science Museum*) 44
Plane-tabling (*Smith Elder & Co*) 48
A fine quality compass (*Science Museum*) 51
Astrolabe (*Science Museum*) 53
A variety of transitional navigational instruments (*Science Museum*) 54
A nocturnal (*Science Museum*) 56
Hadley's octant (*British Cyclopaedia* 1835) 56
Action of reflecting octant, sextant, or quadrant (*Science Museum*) 57
Sextant (*Science Museum*) 57
A diagram of a sextant (*General Astronomy* 1889) 58
An artificial horizon (*Science Museum*) 59
Twentieth-century chronometer (*Dictionary of British Scientific Instruments* 1921) 61
Double sextant (*Dictionary of British Scientific Instruments* 1921) 62
Distance-finder (*Dictionary of British Scientific Instruments* 1921) 65
Augsburg ring dial (*Sotheby & Co*) 68
Selection of Butterfield dials (*Phillips*) 69
Diptych dials, a pillar dial, a pocket sundial, and a tablet dial (*Sotheby & Co*) 70
Armillary sphere (*Phillips*) 72
Study of the heavens (*Magazine of Art*) 76
Galilean telescope (*Science Museum*) 78
Various forms of reflecting telescopes (*General Astronomy* 1889) 79
Transit instrument (*General Astronomy* 1889) 80

LIST OF ILLUSTRATIONS

Mounted telescope (*Science Museum*) 82
Altitude and azimuth instrument (*General Astronomy* 1889) 83
Handbills in both French and English (*Picture Magazine*) 84
Various types of lenses used in telescopes and microscopes (*General Astronomy* 1889) 85
Trade card of Henry Pyefinch (*Science Museum*) 87
Patent for brass draw-tubes (*Science Museum*) 89
Equatorial telescope 1874 (*Graphic*) 90
A refractor (*Science Museum*) 91
Transit instrument and Ramsden's quadrant (*British Cyclopaedia* 1835) 91
Gyroscope (*General Astronomy* 1889) 94
A reflecting telescope (*Phillips*) 96
Transit instrument (*Science Museum*) 97
Equatorial telescope (*Science Museum*) 97
Quadrant (*Science Museum*) 98
Terrestrial and celestial globe (*Science Museum*) 101
Orrery (*British Cyclopaedia* 1835) 102
Orrery (*Phillips*) 103
A globe worked by clockwork (*Science Museum*) 104
Orrery (*British Cyclopaedia* 1835) 105
Trade card of Thomas Wright (*Science Museum*) 106
Dispersion of light and a spectroscope (*British Cyclopaedia* 1835) 111
Spectroscope (*Discoveries and Inventions of the Nineteenth Century*) 111
Spectroscope with a train of prisms (*Discoveries and Inventions of the Nineteenth Century*) 112
Micro-spectroscope (*Discoveries and Inventions of the Nineteenth Century*) 113
Spark spectra (*Discoveries and Inventions of the Nineteenth Century*) 114
Simple microscope with accessories (*Science Museum*) 117
Barrel microscope (*The Microscope and its Revelations* 1891) 119

LIST OF ILLUSTRATIONS

Pocket microscope (*Science Museum*) 120
Hooke microscope (*Micrographia* 1665) 121
A microscope of c1716 (*The Microscope and its Revelations* 1891) 123
Fine adjusting mechanism and accessories for microscopes (*The Microscope and its Revelations* 1891) 124
Double tripod instrument of Culpeper (*Science Museum*) 125
Microscope of John Cuff (*Science Museum*) 126
Microscope of George Adams (*The Microscope and its Revelations* 1891) 127
Tully's achromatic microscope c1826 (*The Microscope and its Revelations* 1891) 129
A Zeiss dissecting microscope (*The Microscope and its Revelations* 1891) 130
An aquarium microscope (*The Microscope and its Revelations* 1891) 132
Victorian binocular microscope (*Phillips*) 132
Barometer (*Illustrated Magazine of Art*) 136
Barometer with a calibrated wheel (*British Cyclopaedia* 1835) 136
Portable barometers (*British Cyclopaedia* 1835) 138
Cistern barometer (*British Cyclopaedia* 1835) 142
Aneroid barometer 143
Barograph (*Science Museum*) 144
A variety of balances (*British Cyclopaedia* 1835) 155
A fine set of drawing instruments (*Phillips*) 160
Trade card of Tuttell (*Science Museum*) 163
Victorian surgical instruments (*British Cyclopaedia* 1835) 165
Victorian amputating saw (*British Cyclopaedia* 1835) 166
Trade card of John Best (*The Connoisseur*) 167
Ophthalmoscope (*Dictionary of British Scientific Instruments* 1921) 168
Actinometer (*General Astronomy* 1889) 199
Apertometer (*The Microscope and its Revelations* 1891) 200
Anemometer (*British Cyclopaedia* 1835) 201
Anemometer (*Illustrated Times*) 202

LIST OF ILLUSTRATIONS

Clinometer (*Dictionary of British Scientific Instruments* 1921) 208
Colorimeter (*Dictionary of British Scientific Instruments* 1921) 209
Azimuth compass (*British Cyclopaedia* 1835) 209
Compass fitted with corrector to counter effects of magnetism (*British Cyclopaedia* 1835) 211
Mining dial (*Dictionary of British Scientific Instruments* 1921) 212
A compressor (*The Microscope and its Revelations* 1891) 213
Dynanometer (*British Cyclopaedia* 1835) 214
Microtome (*The Microscope and its Revelations* 1891) 214
Goniometer 216
Keratometer (*Dictionary of British Scientific Instruments* 1921) 219
Magnetometer (*Science Museum*) 220
Refractometer 229
Seismograph (*Science Museum*) 230
Tachometer (*British Cyclopaedia* 1835) 233
Telemeter (*Science Museum*) 234
Butterfield dial (*Phillips*) 238
English nineteenth century diptych dial (*Phillips*) 239
Sextant (*Science Museum*) 240
Circumferentor (*Science Museum*) 240
Microscope (*Science Museum*) 241
Electrostatic friction machine (*Phillips*) 242
Trade card by John Gilbert (*Science Museum*) 243
Magic lantern (*Science Museum*) 246
Dudley Adams (*Science Museum*) 247

ACKNOWLEDGEMENTS

Among the dealers, collectors and experts I especially thank for their help are Robert Badgery, Peter Bates, Edward Casassa, Colyn Gates, Raymond Head, H. Higby, John Prestige, David Tallis, Graham Webb, and the staff of the Science Museum, London.

INTRODUCTION

THE INSTRUMENT makers were an élite, and in periods when the accent was on extravagance and gorgeous display they continued to produce their functional and beautiful instruments, aloof from the frenzy of progress. In the Victorian age, when bigger and better factories were pouring out a multitude of fantastic and bizarre products, the makers of scientific instruments remained a clique, reluctantly expanding their work forces to cope with the demands of a fast-growing technology.

This is not to say that they were rooted to the past. They were happy to use the machine tools of the new age, the various kinds of lathes and the dividing engines that ensured accurate graduated scales and dials, but it was the workshop and not the factory that was the centre of their life, though the future was anticipated in 1855 when the first telescope factory was opened in the north of England.

On many occasions the progress of instrument making was held up by seemingly insoluble problems. The production of microscopes and telescopes in quantity had to wait until there was a sufficient supply of consistent optical glass, and small precision instruments were handicapped by the absence of accurately threaded screws, essential for the slow-motion mechanism of microscopes. New methods of handling brass formulated in the early nineteenth century revolutionised the manufacture of instruments.

Commerce was the spur to the astonishing wealth of navi-

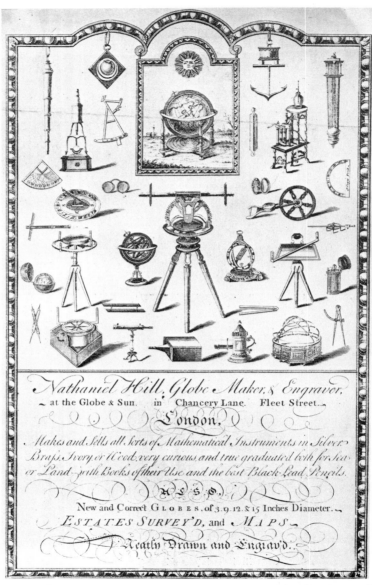

1 The instrument makers were an élite and this is reflected in their beautifully designed trade cards and advertisements.

INTRODUCTION

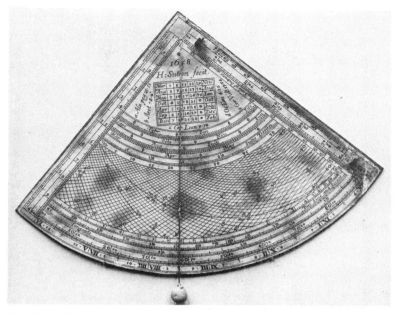

2 An early navigational aid was the simple quadrant, such as this one, made by H. Sutton in 1658.

gational instruments produced. Britain depended on overseas communications, and a merchant navy that could not make its way from point A to point B in the quickest possible time without getting lost was a navy at a disadvantage. The Portuguese mariners used compasses, charts and inefficient astrolabes. The astrolabe disappeared from the nautical scene about 1670 when it was superseded by more efficient tools. In 1731 Hadley's reflecting octant pushed British navigation into a new age, and shortly afterwards John Harrison's chronometer made it possible for the master of a ship to find his position on a chart with unerring accuracy.

Navigational technology was backed by the government. By 1815 £101,000 had been dealt out by the 'Commissioners for the discovery of longitude at sea' and without this body the chronometer would have taken longer to arrive. Surveying was

INTRODUCTION

3 *The theodolite, the most important instrument of surveying.*

also backed by the state. Rough and ready methods of measuring land areas were no longer good enough, and for the assessment of land values a host of sophisticated equipment was devised, including the most important instrument of surveying, the theodolite. The need for the precise measurement of land brought with it the technique of triangulation, using complex

INTRODUCTION

mathematical tables.

Where there was no immediate incentive in terms of cash or commerce innovation and progress languished, and astronomy and microscopy owed their development to rich amateurs who not only commissioned instrument makers to create telescopes and microscopes for them but themselves made their own in their own workshops. A great impetus to instrument making was furnished by the interest of King George III in scientific matters, and the tradition of royal involvement in such things was taken over by the Prince Consort, who actively encouraged the arts and sciences. Alas, this concern was not transmitted to Queen Victoria when Prince Albert died, but by that time it did not matter. The Victorians were avid believers in progress in all spheres, and the production of all kinds of

4 King George III took a great interest in science, and his instrument maker George Adams of Fleet Street, London, enjoyed enormous prestige.

INTRODUCTION

5   *An early Victorian trade card illustrating the immense variety of scientific instruments being manufactured not only in London but in the provinces.*

scientific instruments and gadgets was limited only by the capacity of the instrument makers to fulfil their orders.

The instrument makers were craftsmen in a closed society reminiscent of the guilds of the Middle Ages, and they had a clear idea of what their jobs were. For a considerable time they refused to countenance the notion that thermometers and barometers belonged in their sphere, and thermometer and barometer makers themselves belonged to a rival group, though late in the eighteenth century this distinction was lost and we find many of the famous names on late Georgian thermometers and barometers.

When one considers the wide variety of instruments produced, it is surprising that so little is known of their makers, with certain honourable exceptions. Some makers are known only by their names on the instruments they created, others from trade cards and ephemera. The chronicles of the time did not consider them worthy of perpetuation. They were tradesmen, unfit to be mentioned in the same breath with minor members of the aristocracy.

With the coming of the nineteenth century anonymity went further. Workshops were less cosy, production-flow techniques inspired by the American and British arms factories were brought in, and the practice of farming out components to specialist instrument makers was widespread. A Dollond inscription on a mid-nineteenth century telescope was a trademark and not a guarantee of a custom-made product.

Despite the wonderful mechanics of eighteenth-century instruments and the ingenuity in coping with novel problems displayed by seventeenth-century instruments, there is no question that nineteenth-century instruments are more efficient. The wide use of the achromatic lens patented in the mid-eighteenth century meant that precision instruments of all kinds could be fitted with small efficient microscopes for reading the scales. There was no longer any question about the quality of the materials used; improved methods of processing brass and steel and the mass production of good optical glass by the new industrial giants, such as the Chance Brothers of Birmingham, guaranteed a predictable and efficient instrument. The great advantage of Victorian scientific instruments, which are naturally those most often found in these days when early instruments are often very expensive, is that not only are they beautiful objects but that they work as well as they ever did. Victorian instrument makers did not believe in planned obsolescence.

CHAPTER 1

SURVEYING INSTRUMENTS

SURVEYING INSTRUMENTS are amongst those most widely collected, and although the theodolite and the level are the two basic instruments belonging to this group, surveying encompasses a wide variety of tools, many of which remained basically unchanged for centuries.

Surveying is concerned with finding out and recording surface features of the earth accurately to scale on a map. Of course, there are degrees of accuracy, and in early times an approximation was good enough. The sophistication of surveying instruments went hand in hand with the demand for precision. In the nineteenth century, with such activities as the construction of railway networks, accuracy became of prime importance.

Evidence of some kind of land surveying dates back to at least 3000 BC when the civilisations of Egypt and Babylon demanded the setting down of distances, the division of land into plots, and the construction of a road system. The first requirement of a surveyor is to be able to set out lines at right angles, and perhaps the first of all surveying instruments is the groma, a simple Egyptian device consisting of two small sticks lashed together in the middle to make a cross. From the four ends were suspended small pieces of limestone to act as plumblines. The four lines were made from palm-leaf fibre, and when two of them were used for sighting, the other pair determined

SURVEYING INSTRUMENTS

the direction at right angles. The principle of the plumbline was also used in the construction of buildings.

Linear measurement was similarly rough and ready, and distances were measured using a cord with a knot at regular intervals. The Egyptian and Babylonian instruments were developed by the Greeks, especially Hero of Alexandria (fl 100 BC) and Eratosthenes a century earlier, who brought method into surveying. Parallel techniques were being evolved in China during this period, but it was left to the Romans to systematise surveying, and although they did not introduce any new techniques they streamlined the methods they took over from the Greeks. Their ambitious programme of road building, viaduct and aqueduct construction, and everything involved in colonial expansion was accomplished using relatively primitive equipment, but so efficient was their use of the developments of the Egyptian groma and similar tools that Julius Caesar was able to project a map of the Roman Empire, a task brought to fruition during the reign of Augustus.

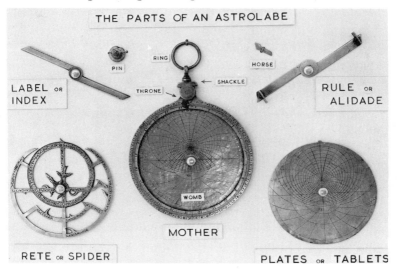

6 *The astrolabe was pressed into service though it was strictly speaking an astronomical and navigational aid.*

SURVEYING INSTRUMENTS

During the Middle Ages there was little encouragement for map-making and surveying, and the instruments used were adapted from navigational and astronomical fields, areas of interest that carried greater prestige than humble surveying, and we find that by the end of the fourteenth century the astrolabe, the quadrant, and the magnetic compass were all being pressed into the service of surveying. Of these, the most interesting and spectacular is the astrolabe (a word derived from the Greek 'to take a star'), which was eventually superseded by the quadrant or sextant. It is defined as an instrument for taking the altitude of a star or heavenly body, and as such belongs to astronomy rather than surveying. Basically it is a calculating device consisting of a circular plate and a rotating rule with sights, though early astrolabes were spherical rather than planispheric.

With such an imprecise instrument as the astrolabe it is not surprising that early maps based on astronomical determinations were inaccurate, though the Arabs seem to have been the most efficient users of it, and their nautical charts were of great assistance not only to them but to western navigators. In 1450 the Arabs were acquainted with the use of the compass and could make charts of the coastline of those countries they visited, and in 1498 Vasco da Gama was shown a chart of the coastline of India mapped out by Arabs, and there is no doubt that such practical aids were of inestimable use to him.

Being of more use, nautical surveying preceded land surveying (apart from small plans) and true land surveying was coincident with an intellectual interest in the size and configuration of the earth by means of exact measurement. The arrival of the science of geodesy encouraged the development of new surveying techniques, the most important of which was triangulation, which arrived in the sixteenth century. The basis of triangulation is a base line of a precise length, and the success of this method of surveying depended on tools for measuring the base line, and tools for measuring angles, the most important of the latter being the theodolite.

SURVEYING INSTRUMENTS

In base-line measurement it is crucial to be absolutely accurate, for an error is magnified by the process of triangulation. Simple land surveys did not call for the same degree of precision, and for these, quite crude methods of determining linear measurement were acceptable. The quickest and most convenient of these methods is based on the principle of series of cogged wheels connected to the hub of a larger wheel, which was pushed over the ground using a handle, a principle that was known to Hero of Alexandria and Vitruvius Pollio, who flourished nearly 2,000 years ago. Both these described a hodometer, which, via eighteenth and nineteenth century variants called waywisers, perambulators, or surveyors' wheels, led to the odometer or mileage indicator, incorporated in the speedometer of a motor car.

Many of these surveyors' wheels are beautifully made and are much collected; some of them are quite large, for the greater the circumference of the wheel the greater the accuracy of the

7 Left. *A mid-eighteenth-century waywiser, sold by Sotheby's in 1972 for £620.*
8 Right. *Waywisers were used well into the nineteenth century for survey work. This Negretti and Zambra instrument was sold in 1972 for £150.*

data fed to the dial, which was frequently set between the wheel and the handle so that the operator had instant information. In the more refined specimens there was an inner and an outer dial, and the distances traversed were expressed in miles, furlongs, poles and yards. It is interesting to note that it was due to the science of surveying that these terms achieved a permanent place in the English language.

The sixteenth century also saw the arrival of the pedometer, a device which was even more convenient than the surveyors' wheels and can be traced to south Germany. The earlier instruments did not provide a reading of the distances walked, only a record of the number of paces taken. Pedometers were quite small, and attached to the wearer's belt; a cord connected with a projecting lever and one of the surveyor's feet, and when he walked the lever operated a pawl which caused a hand on the instrument to rotate. There were usually four dials, operating on the principle of the dials on a gas meter, the first, connected with the projecting lever, registering units, the second 10s, the third 100s, and the fourth 1000s. The earlier pedometers were functional, but during the eighteenth century many beautiful pedometers were sold, made of brass, and decoratively engraved, with dials of white metal. More recent pedometers look like watches, and are activated by a hidden pendulum moving up and down as the user walks. The more sophisticated of these incorporates a device that converts paces to distance, but on all accounts the pedometer can give only an approximation of distance walked and are clearly less accurate than the surveyors' wheels.

No one would have dreamed of using either surveyors' wheels or pedometers for the measurement of a base line to be used in triangulation. In 1620, Edmund Gunter recommended the use of a chain, four poles in length and made up of 100 links. Four poles equals 22 yards, and consequently this length was called a chain. The point about the measuring chain was that it was flexible, and it was not until the early part of the nineteenth century that printed linen tape to some degree supplanted the metal chain. The chain has been described as the land measurer's most important instrument, but it was still not considered accurate enough for triangulation. Surveyors on the continent employed wooden rods from 2 to 5m long, but it was soon discovered that such traditional instruments were subject to changing temperature or humidity. Italian and French surveyors topped their measuring rods with copper or brass, and reduced contact error between the rods by deliberately leaving an interval. These intervals were measured with beam compasses.

Two surveyors were intimately involved in early triangulation procedures, Cassini de Thury (1714–84) and William Roy (1726–90). In his triangulation of 1733–40 de Thury used both 24ft wooden rods, tipped with iron, and iron bars 15ft long. Of these men the most important to British map-making was Major-General Roy, but it was de Thury who proposed in the year before his death that the relative positions of the Royal Observatories of Paris and Greenwich should be determined with the greatest possible accuracy by means of triangulation.

King George III was keen on the project, as it was the Royal Society when the triangulation method was explained to it: by using only a relatively short base line the remainder of the territory could be mapped out by the employment of a continuous network of triangles, only the angles of which needed to be determined.

The principal base line on this side of the English Channel was located on Hounslow Heath and measured 5 miles, and a 100ft steel chain was made for Major-General Roy by Ramsden, the celebrated instrument maker. This was checked by overlapping rods of Riga pine, but these wooden rods proved unstable in varying humidity, and the final measurement was made with glass tubes. Roy was so impressed with the performance of the Ramsden chain — made of steel strips joined as in the manner of a folding steel rule — that he employed it again in 1787 for the Romney Marsh base.

Many attempts were made by scientists to create a measuring

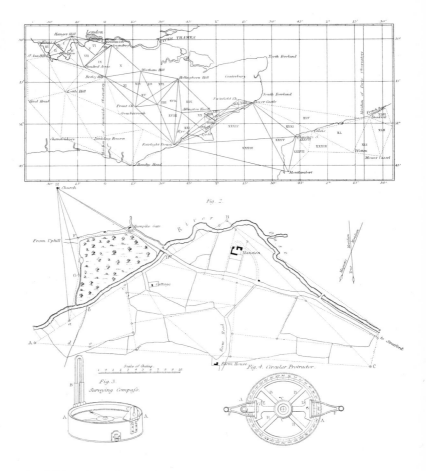

9 This plan shows the pattern of triangulation carried out in south-east England, together with the triangulation of a small area and two useful surveying instruments, the surveying compass and a protractor.

tool that was not subject to the vagaries of the atmosphere. In 1792 the Frenchman J. C. de Borda used bars of two metals (copper and platinum) and a Prussian scientist involved in the triangulation of East Prussia used iron and zinc. For the triangulation of Ireland in 1827–8, the director of the Ordnance Survey invented an apparatus in which two parallel bars of brass and iron were connected at their centres and linked at the ends by pivoted metal tongues, and as these double rods had an error of only 1 : 200,000 they were used in the English survey twenty years later. The accuracy and precision of the Ordnance Survey maps of this period owe much to the diligence of scientists and surveyors in attempting to obtain the perfect measuring rod. As scientific instruments such rods and chains are hardly exciting, but they illustrate an interesting aspect of scientific endeavour.

Roy's was the first accurate triangulation to be made in Britain, and in 1791 the Master-General of the Ordnance began preparations for a map of Britain on the scale of one mile to an inch (1 : 63,360), the scale that is used today. Roy's base line measured on Hounslow Heath had carried an error of only 1 : 158,000, a mere 2in over the 5 miles base line. Ramsden had provided the steel chains for the 1787 attempt to ascertain the difference in latitude and longitude between the observatories

10   A Ramsden theodolite, of a type very highly regarded by collectors.

of Paris and London, and he was also called upon to provide a 3ft theodolite. In triangulation there are the 'great triangles', and for this the 3ft theodolite was used (it took three years to make); for secondary triangles smaller theodolites were used, made by Ramsden and by the great instrument makers Troughton and Simms.

The accuracy of the theodolite is as important to the success of a triangulation project as the correct measurement of a base line; if the angles are not measured accurately, errors multiply. By 1825 the whole of England and Wales had been triangulated and all but the six northern counties definitively mapped. The triangulation of Scotland was well nigh complete. All this was not merely an academic exercise, for maps were increasingly necessary as a basis for land valuation, and when a survey of Ireland was ordered in 1825 it was by Parliamentary decree for this very purpose.

A number of talented surveyors had made their appearance in the seventeenth century, including two Englishmen, John Norden and Aaron Rathbone. Norden introduced the circumferentor into English surveying, though its inventor seems to have been Gemma Frisius who lived in the first half of the sixteenth century. The circumferentor consists of a graduated circle of degrees over which moves an alidade. What is known as the Holland circle type had four sights (separated by 90°) on the main circular plate. These instruments, extremely collectable, vary a good deal in ornamentation, and sometimes the interior plate, moving around inside the circle marked out in degrees, is cunningly shaped and ornately engraved. With the aid of the sights and the circular scales the instrument could be used to measure either horizontal angles, when it was mounted horizontally on a stand, or vertical angles, when it was suspended from a shackle.

A simple instrument that in some ways resembles the Holland circle is the cross staff. In surveying it is vital to measure right angles, and this was the job of the cross staff. There are two basic kinds of cross staff. One consists of an octagonal brass box

11 A fine circumferentor by Macarius dated 1676.

with slits cut in each face so that the opposite pairs form sight lines. The other consists of four arms fixed at right angles each to each, with a slit at the end of each arm. The cross staff usually has a hollow stem so that it can be mounted on a rod.

An instrument that does the same thing is the optical square; there are two kinds, one using two mirrors, the other a prism. In essence the optical square is a small metal box containing two mirrors A and B, the former being completely silvered, the latter silvered in the lower half only, the upper part being transparent. The eye is placed at the slit C, and a pole F is viewed through the transparent part of B and the opening D. Now if there is a pole at E it will be reflected by A on to the silvered part of B, and thence reflected to the eye. The angles of the mirrors are adjusted so that when the pole F appears vertically over the reflection of pole E, the two lines are at right angles.

SURVEYING INSTRUMENTS

12   A sturdy cross staff by Dollond.

13   A mid-eighteenth-century brass surveying instrument by Benjamin Martin, sold in 1971 for £380.

The principle is that of the sextant and other angle-measuring instruments using the system of lining-up the images on two mirrors.

The prismatic optical square uses a pentagonal prism, cut so that two faces contain an angle of 45°. It is used in the same way as the mirror optical square. Cross staffs and optical squares are the latter-day gromas.

In terms of collectability such minor instruments are not in the same class as the circumferentor, a beautiful transitional tool between the astrolabe, which it superficially resembles, and the more sophisticated angle-measuring instruments. Later surveyors found that when examining early county maps relying on circumferentors and similar tools there were errors of nearly three miles in eighteen, and it was difficult to use the circumferentor without an error of at least 2° creeping in.

In instruments that rely entirely on the unassisted eye there is bound to be a degree of error of this order, though greater precision was obtained when the Vernier scale was introduced.

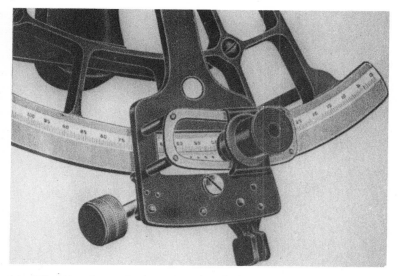

14 The Vernier scale, introduced to help establish minute sections of the arc.

Pierre Vernier (c1580–1637) is credited with the invention of the auxiliary scale in 1631, and the purpose of this short moveable device, sliding over the principal graduation scale, was to split already small sections of the scale into intervals yet smaller. The ability to establish minute sections of the arc brought a new dimension into surveying, though to some extent the advantage was lost, as circumferentors relied entirely on the naked eye.

There are many surveyors' instruments that derive from the circumferentor or are variants of it. Amongst the most important are graphometers. In 1597 details of the graphometer were published; this was a circumferentor with a semi-circle scale

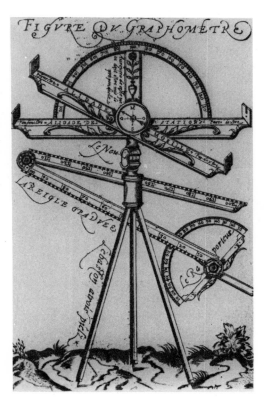

15 The graphometer was first introduced in France in 1597, a circumferentor with a semi-circular scale.

instead of a complete circle, and because it was fairly easy to make and use it was popular amongst the surveying fraternity, especially in France, and although the circumferentor became obsolete in the eighteenth century the graphometer was used well into the nineteenth, even though theodolites were then in universal use. In many graphometers the semi-circular scale could be detached from the body of the instrument and used as a protractor for plotting angles in the field.

The surveyor's quadrant had a scale of only a quarter-circle and so was, in effect, half a graphometer or a quarter of a

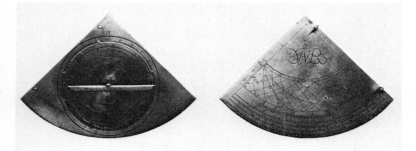

16   A simple quadrant, used for a variety of purposes.

circumferentor. In its earliest form it was used in astronomy to find the altitudes of stars and planets, and the quadrant was adopted by surveyors to ascertain the heights of terrestial subjects. Naturally a *quart de cercle,* as it was known on the continent, was sufficient for this, and was smaller and more portable than its predecessors. By measuring the base distance to, for instance, a cathedral, then taking a reading from the top of the spire with the quadrant, the height of the cathedral can be read off from a mathematical set of tables; many of these tables were glued inside the lids of the cases of quadrants and other portable angle-finding instruments. It is interesting to note that many fine seventeenth-century surveyor's quadrants doubled as artillery quadrants, with appropriate scales for use in gunnery on the reverse.

Circumferentors, graphometers, quadrants — each half the size of its predecessor — it is not surprising that the quadrant itself was halved, resulting in the octant; this was a much rarer

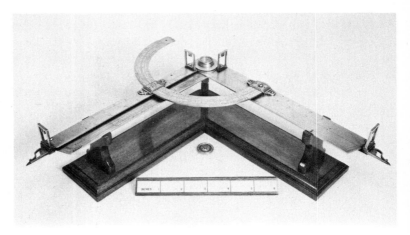

17 A graphometer of Italian design. It is interesting to compare this with the 1597 graphometer.

breed than the quadrant, for obviously it had less utility. Of all these instruments perhaps the circumferentor was the most significant, and it was this instrument that led to the 'common' theodolite, almost indistinguishable from the circumferentor.

A technique widely employed by surveyors in the field, and especially military engineers engaged in surveys of fortifications, was plane tabling. Plane tabling was by no means a substitute for the more intricate procedures, but was useful for making rough notes or for preliminary work. English authorities of the seventeenth and eighteenth century considered that plane tabling should be confined to 'Townships and small Inclosures'. The equipment needed for plane tabling consists of a drawing board, the lower edge of which has a joint for attachment to a stand (to give rigidity in the field), and a brass alidade or sight-rule — a straight edge fitted with sights. Unchanged in essentials since it first made its appearance in the early seventeenth century, the use of the alidade enabled the surveyors to record directly on to paper their findings and their chartings.

There are two kinds of plane-table surveying — the production of complete maps, and for filling in detail when the main

18 The sturdy kind of compass used in rough conditions.

work has been done by theodolite. Plane-table surveying was used extensively on Scott's antarctic expeditions, and large areas were mapped without using other equipment. Sights are taken on to an object with the table at two separate points but with the same directional alignment. Rays are drawn on the board along these sight lines, and these must intersect at the plotted position of the object. The two stations are a known distance apart, and as the angles between the stations and the object are known, as the board has been orientated by means of a compass, the distance of the object is known, and from the data in possession of the surveyor further triangulation procedures can be carried out. Briefly it is triangulation carried out using the simplest methods.

Plane table surveying is also used involving one station, set in the middle of the area to be surveyed, and radiating lines are drawn to known points.

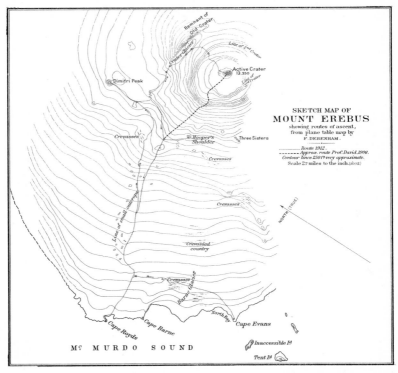

19 Plane-tabling could result in first-rate maps, as can be seen here, drawn up during the 1912 Antarctic expedition.

A mathematical instrument also used in surveying was the sector. This consisted of two equal flat arms hinged together as in a joint rule. The arms bore various scales, which were used to solve problems not only in surveying, but in geometry and gunnery as well. The French gave the sector the name *compas de proportion,* a good deal more explicit than its English title. To divide a given line into the same proportions as a scale on the sector, the sector was opened until the ends of the two arms

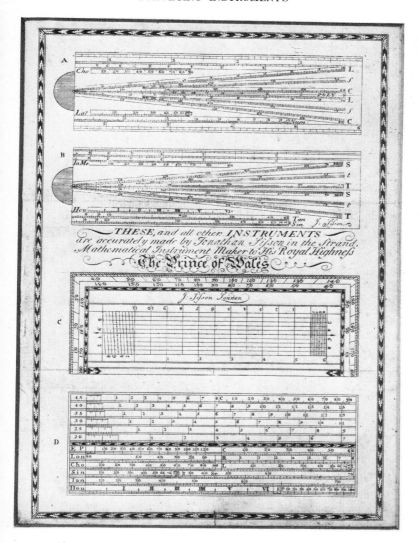

20 A variety of sectors, illustrated in a Sisson trade card.

were as far apart as the length of the line; verticals were then dropped from the scale to the line. Sectors are extremely collectable, for there is a good deal of delicate engraving with the scales; they are found in brass, bone, wood and ivory.

It was increasingly found that the unaided eye was unable to cope with the more delicate work demanded of surveyors. In the first half of the eighteenth century Thomas Wright (1711–86) and Jonathan Sisson (1694?–1749) replaced the open sights of the common theodolite by a telescope with cross-hairs; this was mounted on a graduated arc, so that the surveyor could observe horizontal and vertical angles simultaneously. It was logical to construct an instrument that combined the functions of measuring altitude and azimuth (horizontal angles). The altazimuth theodolite — altazimuth being an ugly portmanteau word combining altitude and azimuth — was the complete answer to most surveyors' dreams. There were other factors other than the adding of a telescope that made the altazimuth theodolite the principal instrument used in triangulation. Improvements in the design and construction of angle-measuring instruments meant that optical accuracy was increased, that the circles and the alidades were mechanically true — improved tool-making techniques were now guaranteeing perfect circles and straight lines — and that there was now a precise division of the graduated scales. For the first time the Vernier scale could come into its own. Although a compass was still part of the theodolite, it was now of less importance; of much more consequence was the incorporation into the design of a spirit level, a device invented by Thevénot in 1666.

Hero of Alexandria and Vitruvius Pollio had described water levels, using still water as a basis for determining horizontal directions, and the spirit level was a variation of this, an application of the everyday fact that the action of gravity makes the surface of a still liquid a horizontal plane. By filling a glass tube with alcohol or ether and leaving a space for the bubble, observation of the bubble made it much easier to determine the horizontal, and replaced the rough and ready method of the plumb line, from which a horizontal was calculated. Levels incorporating a telescope can be large and elaborate, and are much collected.

SURVEYING INSTRUMENTS

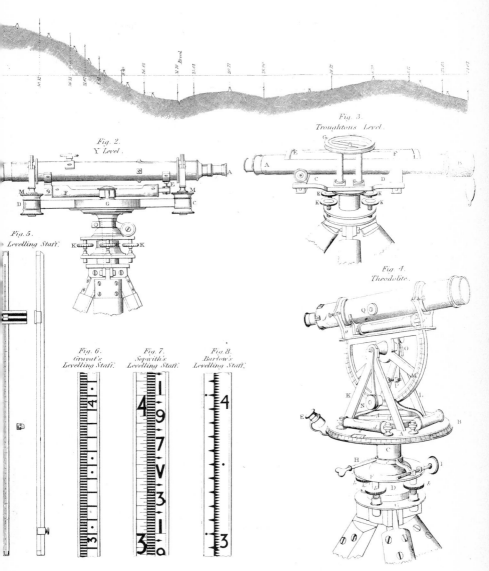

21 The illustration shows an early nineteenth-century theodolite, and two levels, together with the levelling staffs used in conjunction with these instruments.

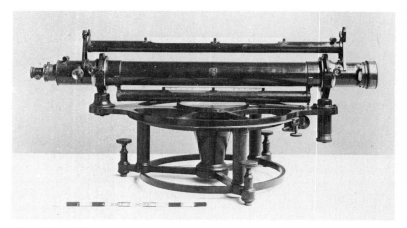

22　*A level incorporating a telescope was a logical step forward in surveying, and, previously undervalued, levels are now becoming highly desirable.*

Mention has already been made of the triangulation project of Major-General Roy when he linked Paris and London with his network, and the search for true base line. No matter how accurate the base line was it would have been of little account without a first-rate theodolite. It was common sense that the larger the theodolite the smaller the amount of error, and Ramsden's great theodolite had a 3ft horizontal circle and weighed 200lb. Just as half a century earlier the instrument maker Sisson had first successfully combined theodolite and telescope, so Ramsden in 1787 incorporated the most modern knowledge available in instrument making. Angles could be measured within one second of arc, and readings could be made up to 100 miles. The achromatic lens introduced by John Dollond (1706–61) in 1758 permitted shorter refracting telescopes to be made, John Bird (1709–76) introduced a means of graduating astronomical quadrants more accurately, and between 1766 and 1775 Jesse Ramsden invented what were known as circular dividing engines. All these contributed to the success of the great theodolite.

SURVEYING INSTRUMENTS

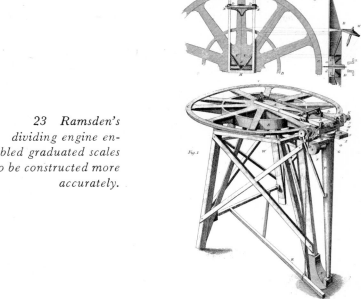

23 Ramsden's dividing engine enabled graduated scales to be constructed more accurately.

Precision mechanics played an immense part in the evolution of scientific instruments. Inventors were obliged to keep up with the pace set by their contemporaries in other fields. On the surface it might seem strange that telescopes were not allied with angle-measuring instruments such as the quadrant or the common theodolite at an early date, for obviously telescopes were far superior to ordinary sights. But until telescopes were of a reasonable size it was ludicrous to couple them with surveying instruments much smaller. Until the arrival of 'Dumpy' level in 1848 levels were very cumbersome. It must be remembered that surveying tools were outdoor objects, and that whereas an astronomical telescope could by its nature be unwieldy this would not be true of telescopes built into the structure of surveying instruments. The 1758 achromatic lens enabled manageable telescopes to be made.

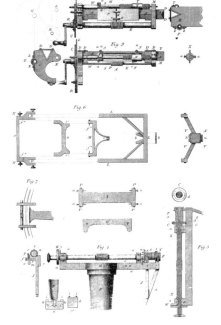

24 It seems incredible that this complicated gearing and machinery should have been made by Ramsden before the start of the nineteenth century.

25 Achromatic lenses, patented in 1758, enabled smaller telescopes to be fitted to surveying instruments, as witness this Troughton and Simms theodolite.

Instrument making derived from clock making, the tools of which were used or adapted, and it is interesting to note that the first great instrument maker of the eighteenth century, George Graham (1673–1751) was the pupil of the genius of clock making, Thomas Tompion. One of the most important operations in the construction of sophisticated angle-measuring instruments is the cutting of an accurately threaded screw. In 1770 Ramsden constructed two screw-cutting lathes which, for the first time, were wholly successful. The dearth of accurately threaded screws also held back the development of microscopes, for with the tiny screws used in the slow motion of microscopes there was no alternative to cutting them on a lathe.

More important still than accurate screws for the surveying instruments were accurate scales. The increasing precision of theodolites and the use of the Vernier scale made old methods of dividing the graduated limbs of instruments and rules obsolete, including the method using transversals, in which tenths were divided into hundreds by taking a diagonal down each tenth. On circular scales, such as those on circumferentors, the instrument makers used dividers and a set of tables. It was clear that although these methods were adequate to a degree, they lent themselves to inaccuracy. There was also the question of time. The demands of navigators and surveyors multiplied; British instrument making was supreme in the world.

It was clear that the division of graduated limbs must be done mechanically, with no room for operative error, and division engines were introduced into the workshop, deriving from the clock makers. An early attempt by the English clock maker Henry Hindley in 1739 was still-born, shortly afterwards the Duc de Chaulnes invented two dividing machines, one for circles, the other for straight lines, and he was followed by Ramsden. By using an accurate master plate, the dividing engine turned out consistent replicas of true scales, guaranteeing accuracy.

Although quadrants were the favourite instruments of surveyors until about 1780, it was recognised that by their very simplicity they could provide faulty readings. When a surveyor is measuring angles he does it not once, but several times, checking the original readings. It is therefore possible to make the same

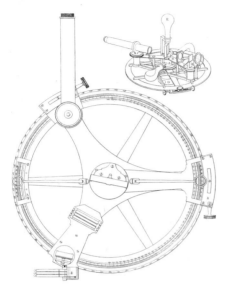

26 *Diagrams of Dollond's reflecting circle and Troughton's reflecting circle.*

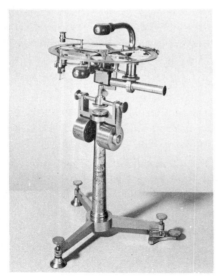

27 *It is interesting to compare this photograph of a Troughton reflecting circle with the diagram.*

erroneous readings time and time again. In 1752 J. T. Mayer (1723–62) considered using a small circle and repeating the measurement of the same angle several times without reverting to zero between each measurement. Any error of observation was therefore rendered negligible by the number of observations. (An angle of 57.6° may be read as 58° on each of several occasions, whereas if the cumulative method is adopted, and there is no return to zero, the sum of five readings is 288°, which is then divided by 5.)

Nothing was done for twenty years, when the navigator J. C. de Borda (1733–99) modified the device. Division machines were used to calibrate the circle, which was made in one piece, and although Borda's reflecting circle — it incorporated a telescope and a mirror and was logically a development of Hadley's octant — was considered first as an astronomical aid, surveyors also took it up. The repeating circle was invented by Etienne Lenoir to help in the Paris–Greenwich triangulation of 1788, and was used instead of a theodolite.

Two telescopes were fixed on either side of a calibrated circle, moveable in the plane of the circle, and passing over a frame, elements of which acted as alidades; each alidade had Verniers, read by small microscopes. The repetition of angles using different arcs of the circle cut down observation error. A flexible unit between the base and the circle made it possible to use the repeating circle vertically and horizontally. This instrument resembles the theodolite, but whereas in the theodolite the graduated circles fixed to the centre of the telescope are a minor part of the design, in the French instrument the circle is the predominant feature.

The circle was therefore an alternative to the theodolite, but the British preferred the theodolite which became the standard surveying instrument of the Victorian age. Few of the very large Ramsden-type theodolites were built, but so excellent was his three-foot instrument that it was used from 1787 to 1853. Reflecting circles, repeating circles, and theodolites were all equipped with a variety of scales and adjusting devices. It was crucial to position them absolutely horizontally. The favourite British method was by using four foot screws, though the 'Everest' theodolites made for the survey of India used three. A valuable refinement was the addition of a small mirror to the spirit levels so that the observer could check the position of his instrument without having to move from his observation position. A further refinement, introduced towards the close of the nineteenth century, was the gradiometer, a level fitted with a graduated micrometer tilting-screw beneath the eyepiece.

Theodolites altered little over the last fifty years of the nineteenth century, though considerable stylistic changes have occurred since. Present-day theodolites are much more compact than their predecessors, and can weigh less than ten pounds. The metal scales have been largely replaced by illuminated engraved scales, and matt-painted surfaces have replaced brass. Compact easily carried theodolites were a boon to the members of Scott's Antarctic expedition.

Theodolites were immensely useful in the checking of data; in the mapping of new country they were invaluable. The need for such instruments was especially pronounced in America, and in 1753 John Lewis of New York announced 'a New Method of Navigation . . . an excellent Method of Trigonometry here particularly applied to Navigation'. The casual method of dealing out land rights and inaccurate surveys made the use of modern methods imperative. The demand for information led to the wide sales of books on surveying, such as *An Accurate System of Surveying* (1796) and *A Compendious System of Practical Surveying* (1799).

The accent was on the practical, for in a country that was short of specialists the material had to be presented in a form that could be assimilated by a layman turned surveyor. The country was not only short of surveyors and cartographers, but of instrument makers, and the most unlikely people turned to instrument making, such as Rowland Houghton, described variously as a commissioner, proprietors' clerk, and collector, who in 1737 announced in the Boston press that he had 'lately improv'd on his new Theodolite, by which the Art of Surveying is rendered more plain & easy than heretofore'.

Because of the demand for surveying instruments many watch- and clockmakers turned their attention to instrument making; one such was James Jacks of Philadelphia, who made and sold not only clocks and watches, but theodolites, surveyors' compasses, quadrants, fishing rods and reels, billiard balls, and whips. The theodolites made by native Americans were rough and ready, but were none the less efficient. There are not many of them extant, and their rarity both then and now can be judged by the fact that most instruments are equipped with a provenance. The theodolite used by Orange Warner Ellis about 1780 for surveying the border between the United States and Canada is typical of the instruments made at this time — sturdy, no nonsense, and unsigned.

In most cases the precision of a theodolite was not really necessary, and surveying compasses were much more useful for on the spot surveying. These varied from fine instruments well up to the standard of English instrument makers to compasses carved from hickory, maple, cherry, walnut, birch, pine or oak, indeed any wood that happened to be handy. The emphasis on wood was due partly to convenience, but also to the influx of German and Dutch makers of clocks with the movements made of wood, and to the background of instrument makers as cabinet makers and carpenters.

Some of the compasses made from wood are beautifully made and are certainly not the work of backwoodsmen. Makers of brass compasses also made wooden ones. These include Daniel Burnap, Samuel Thaxter, Thomas Greenough, and John Dupee, and the records of Burnap indicate that wooden surveying compasses were made for non-professional surveyors who wanted a reasonably accurate cheap compass — his brass compasses sold at £4 – £6, his wooden ones at £2.

Other semi-professional surveying instruments surviving from eighteenth-century America include graphometers and similar angle-measuring instruments. There are certain especially interesting instruments, such as a type of quadrant 'invented by P. Merrill, made by John Kennard, Newmarket', which have been constructed to suit the exigencies of the moment, perhaps made from engravings of European instruments or using the information supplied by the early American primers on surveying. Plane tabling, one of the oldest methods of surveying, using an alidade and a chart, was no longer used in Britain except for rough

28 *Plane-tabling was still occasionally used, and proved invaluable in Scott's exploration of Antarctica in 1912.*

plans and small jobs, but it is clear from surviving American instruments that this method continued to be widely used in circumstances where nothing better was available.

It is not surprising that many British instruments found their way to the United States, including pieces by George Adams, the prestige instrument maker, supplier of instruments to King

George III, W. and S. Jones, and Benjamin Cole, and it is to the credit of the Americans that instruments by their best instrument makers, the Rittenhouse brothers and Andrew Ellicott, vie with the products of the great London makers. In particular a theodolite made to his own design by Ellicott in 1789 would be willingly acknowledged by any of the London makers as worthy to rank alongside their own theodolites.

During the nineteenth century, America quite caught up with European instrument makers, and were in the van when photography was utilised in the making of maps, following hard on the heels of the French who made a photographic survey of a village near Versailles as early as 1851. Photographic surveys were carried out using a camera mounted on a theodolite base, with the telescope and the vertical arc on top. Balloons carried the equiprnent and the surveyor to the desired altitude.

CHAPTER 2

NAVIGATIONAL INSTRUMENTS

AIDS TO navigation can be divided into two classes — those carried on board ship such as charts, compasses, sextants, chronometers, logs, and lead lines for determining the depth of the water, and those set up on shore. Whereas there was little state promotion of surveying instruments until the last years of the eighteenth century when the Ordnance Survey was mooted, navigation aids were eagerly sought after, and a good deal of money went into investigation. Britain depended on overseas trade, and overseas trade depended on good maritime communications.

Up to the time of the Portuguese adventurers backed by Prince Henry the Navigator who discovered the Azores in 1419, the rediscovery of the Cape Verde Islands in 1447 and of Sierra Leone in 1460, navigation had been carried out on a free and easy basis, and where sailors left the coast-line they relied on a compass and a chart. The discovery that a lodestone, or piece of iron that has been touched by a lodestone will align itself in a north–south axis has been attributed to, among others, the Chinese, the Arabs, the Greeks, and the Etruscans. The properties of the lodestone were certainly known to the Chinese in AD 121. It does not appear that the compass was in use amongst Europeans before 1400.

The compass must be one of the most familiar of all scientific instruments, and consists of five principal components — the card, the needles, the bowl, a jewelled cap, and the pivot. In the mariner's compass the card is above the needles, though the placing of the card at the bottom of the box was practised by Nuremberg and Bruges compass makers about 1600. The card, or 'fly', was formerly made of cardboard, but later by mica faced with paper, or paper alone. To confuse matters still further the combination of card, needles and cap is termed the *card*, or the *rose*. There are never less than two needles in a mariner's compass. The Admiralty Compass which came into being after the Admiralty had appointed a committee to look into the whole question of compasses in 1837 has four needles. The modern

29 *A fine-quality compass.*

compass dates from 1876 when Sir William Thomson used eight light needles secured to silk threads radiating from a pivot (a small inverted aluminium cup with a sapphire crown) to the rim of the compass. The use of more than one needle was to overcome moments of inertia. The lighter the card (in the sense of fly, needles and cap) the more sensitive the compass is. Thomson's card weighed 180 grains compared with the 1,525 grains of the Admiralty card.

From its inception about 1400 the compass did not alter

much for more than 300 years. In 1616 Barlowe in his *Magnetical Advertisements* complained that 'the Compasse needle, being the most admirable and usefull instrument in the whole world, is bungerly and absurdly contrived'. In 1750 Gowin Knight (1713–72) showed how the magnetic flux of compass needles could be increased, but his 'improved' compass did not attempt to cope with moments of inertia, nor did he cope with the three major problems that occupied the attentions of marine instrument makers (a) the north on a compass fly is not true north (b) high seas and inclement conditions played havoc with the compass (c) iron and steel in the ships caused deviation. With the coming of iron ships the latter problem became critical. Between 1800 and 1803 Commander Matthew Flinders carried out experiments, and found the means of correcting needle deviation by Flinders bars, a combination of magnets and iron strategically placed relative to the compass.

The effect of rough weather was countered by the development of the liquid compass, suggested in 1779 and used by the Danes in 1830. The card nearly floats in a bowl filled with distilled water: 35 per cent of alcohol is added to prevent freezing. An expansion chamber is provided to cope with the expansion or contraction of the liquid. Liquid compasses were fitted with gimbal rings to keep the bowl and card horizontal under all circumstances, and although liquid compasses were supplied to the Royal Navy in 1845, it was not until 1906 that all Royal Navy vessels were equipped with the liquid compass.

Until about 1820 mariner's compasses were mounted on a tripod and moved about the ship as convenient, but later they were mounted on a permanent pillar of wood or brass, the binnacle. The Thomson compass on a rigid platform and provided with the gimbal suspension incorporating springs proved extremely efficient, and is still in use in older merchant ships, though most have gone over to the modern gyro-compass, which first appeared in Germany in 1906.

The compass card is a good deal older than the compass itself; the eight principal points were originally marked with the eight principal winds. The north point was an arrow-head combined with the letter T for Tramontano, one of the winds, and about 1492 this developed into a *fleur de lis,* a convention that is still with us. The east was marked with a cross, and this persisted in British compasses until about 1700. It may be that the use of symbols derives from Arab compasses, which were orientated to Mecca with principal points relating to other places.

The fact that the north of a compass is not the true north was noted by either Columbus or Cabot about 1490. It was also known that this deviation varied. The extent of the variation was found out either by taking a bearing of the pole star,

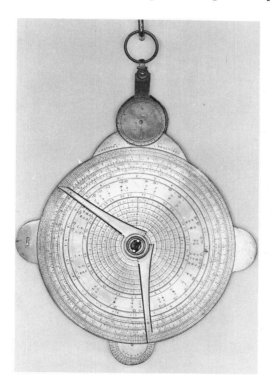

30 *An astrolabe was one of the few instruments available to the Portuguese navigators of Columbus's day.*

or by taking the mean of the compass bearings of the rising and setting sun. The few instruments that Columbus and his contemporaries had could, at a pinch, find out the latitude of their vessels, but the mariners found it next to impossible to discover the longitude. Their customary navigational equipment consisted of a compass, a cross staff or astrolabe, a table of the sun's declination, a correction for the altitude of the pole star, and a chart of sorts, plus a wooden board which folded like a book, the log board, later abbreviated to the log. The speed of the ship was ascertained by throwing over the side a piece of wood on a line, which would drift past two observers posted a known distance apart.

The cross staff was taken over by mariners from astronomy; its job was to measure altitudes at sea, and consisted of two battens, one 36 in long, on which the second batten, a little

31 A variety of transitional navigational instruments including a simple sextant, a compass, a cross staff, and an hour glass.

over 26in long, was fitted at right angles and made to slide up and down. Sights were provided on the second batten precisely 26in apart. At the end of the first batten, which was equipped with a scale, was another sight. By aligning the sights, and moving the scaled batten up and down, angles could be measured which would assist in determining the position of the sun.

The cross staff was used facing the sun, and could thus be both inconvenient and dangerous to the observers' eyes, and to remedy this Captain John Davis invented in 1590 the back staff, which was used by standing with one's back to the sun.

The mariner's astrolabe (more correctly an astrolage) was also taken over from astronomy. It was a simplification of the astrolabe, and consisted essentially of a heavy pierced disc with a suspension ring. Around the edge of the rim were engraved scales of degrees, and the instrument was provided with an alidade. The astrolage was used extensively between 1540 and 1670 when it was rendered obsolete by more accurate equipment. They were made of copper or tin, were a quarter of an inch thick and six or seven inches in diameter.

The nocturnal was used in association with the cross staff or astrolage to find both the latitude and the time of night, using the position of two stars in the constellation known as the 'Little Bear'. The nocturnal consists of two concentric circular plates, the outer one divided into twelve equal parts representing the twelve months of the year, each subdivided into groups of five days. The inner circle was divided into twenty-four hours, each subdivided into quarters. The nocturnal was equipped with a handle. The small disc is set to the date, the pole star is sighted through a hole in the centre of the double discs, and the index arm attached to the common centre of the discs is moved round until the edge of the arm just touches the two stars, known as the Fore and Hind guards. From a set of tables known as the 'Regiment of the Pole Star' the position of these two 'guards' vis-à-vis the pole star could establish the vessel's latitude.

These rough and ready methods, ingenious as they are, had to suffice until more sophisticated instruments were made. The

NAVIGATIONAL INSTRUMENTS

32 *A nocturnal dated 1543.*

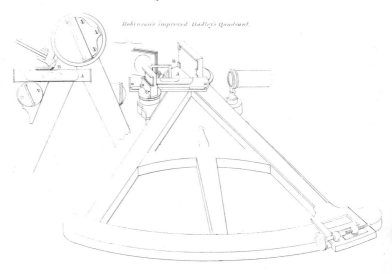

33 *Hadley's octant was introduced in 1731. This was a development of it.*

NAVIGATIONAL INSTRUMENTS

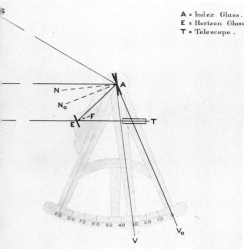
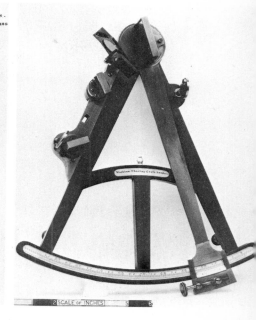

DIAGRAM TO EXPLAIN THE ACTION
OF A
REFLECTING OCTANT, SEXTANT, ETC.

34 Left. *A diagram to explain the action of a reflecting octant, sextant, or quadrant.*
35 Right. *A fine sextant made by Watkins of Charing Cross.*

most significant barrier to progress was the absence of an effective time-piece.

A great breakthrough occurred in the 1730s. In 1731 Hadley introduced his reflecting octant. Until then all instruments at sea for measuring angles either depended on a plumb line or required the user to look in two directions at once. Hadley solved the problem with mirrors. His octant, sometimes confusingly known as a quadrant or navigator's sextant, consists of a sector of 45°. The true sextant is a sector of 60°.

There are two versions of the sextant, the nautical and the box. The nautical sextant consists of a 60° arc, the periphery of which is calibrated. A moving alidade carries a small mirror at its top, the bottom of which moves across the calibrations. On

one of the arms is a sighting telescope, on the other a second
fixed mirror. The height of the star is measured by noting the
position of the alidade when the image of the star, reflected
from pivoted to fixed mirror, thence into telescope, coincides
with the horizon seen through an unsilvered portion of the fixed
mirror.

The box sextant is a miniaturised version of the standard
nautical sextant, and is held in one hand by a projecting handle.

The use of the sextant, the generic name for nautical angle-
measuring instruments, greatly facilitated Cook's surveys in the
Pacific. In his circumnavigation of New Zealand in 1769–70 he

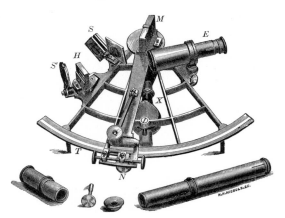

36   *A diagram of a sextant.*

charted 2,400 miles of coast-line in six months, made possible
by the reflecting sextant which became the mariner's favourite
instrument, and could be used for not only vertical but horizontal work. A great advantage was that it was portable, and
could be carried up masts to extend the field of observation.
Sextants, made of metal, were in use well into the closing years
of the nineteenth century. True octants were less favoured,
because the arc was not enough for observations of the moon.

The sextant was also associated with another venture into
the unknown — the artificial horizon. In bad conditions where

NAVIGATIONAL INSTRUMENTS

37 *An artificial horizon, used when the horizon was obscured by fog or other causes.*

the horizon is obscured it is obviously useful to have an alternative. Shortly after Hadley's octant appeared, John Elton suggested fitting two spirit levels to the instrument to enable the user to hold it horizontal, but this did not prove suitable, nor did an ingenious successor, a spinning-top with a glass upper surface. A hundred years later another attempt was made, using the principle of the pendulum. The method most used is an enclosed level using mercury, a method dating from 1812.

In 1714 a body was thought up called the 'Commissioners for the discovery of longitude at sea' which had £2,000 to dispense a year to likely discoverers and inventors. Navigators had the equipment and the knowledge codified in complex tables to find latitudes, but they were still stumped by the longitude. The value of logarithms as a navigational aid had been discovered as long ago as 1614, and Addison's *Arithmetical Navigation* with logarithms for numbers from 1 to 10,000, which came out in 1625, retained its grasp on the nautical fraternity for a century or more. Another theoretical-knowledge gap was filled in 1637 when Richard Norwood, a sailor and a mathematician, got to grips with the problem of the nautical mile, in the end coming up with 2,040yd — 12yd too much.

The challenge of the Commissioners for the discovery of longitude was not met at once, and the first grant was made 23 years after its inception. By 1815 £101,000 had been meted out, and although the board continued until 1828, by that time the problem ceased to intrigue. Although the board was occasionally swept away by ill-considered bequests to eccentrics, it did do an immense amount of good, backing Cook's explorations and encouraging the improvement in timepieces.

By using the tables a navigator could discover the local time, but he could not ascertain his exact position at sea without knowing the time at zero meridian. How could a timepiece be made which would not be affected by the movement of a ship? Clock-makers suggested ways of suspending the pendulum, or of using double balance wheels. The first truly successful timepiece that could be used at sea was the chronometer by the English clock-maker John Harrison (1693–1776), completed in 1735, but it was not until 1759 that his fourth chronometer was fully tested by the Commissioners for the discovery of longitude, and only when a replica had been satisfactorily made and tested by another clock-maker did the commissioners give the whole of the prize that was due to Harrison — £20,000.

Harrison's great invention was the principle of compensation through the unequal contraction of two metals, and the result was the gridiron pendulum. There was a good deal of heated controversy between Harrison and the Astronomer Royal about the value of Harrison's chronometer, and this ill-feeling persisted between the Astronomer Royal and other clock-makers, such as Thomas Mudge, John Arnold, and Thomas Earnshaw, who were also interested in getting a share of the money paid by the Commissioner for the discovery of longitude. By 1805 Arnold and Earnshaw had been paid £3,000 each by the board, though the most important steps to the perfect chronometer (a name invented by Arnold) were made by a French watch-maker, Le Roy, who unfortunately lost interest before it was clear that his innovation (the escapement leaving the balance wheel free during the greater part of its movement) was the key to the future. Although Le Roy (sometimes spelled Leroy) foresaw the principles on which later marine chronometers were constructed, a quarrel with his rival Ferdinand Berthoud encouraged him to discontinue work on chronometers, though Berthoud (1727–1807) went on to become an important figure in the development of the chronometer, introducing two circular balance wheels oscillating in opposing directions and connected by means of a toothed wheel and utilising the gridiron pendulum invented by Harrison. The long time it took Harrison to get his chronometer accepted enabled the French makers to get into the lead. Berthoud in particular was a great innovator, dropped the gridiron method, and finally returned to what the Victorians called 'the power of spring'.

His nephew Pierre-Louis Berthoud (1754–1813) carried on his work, and his four helpers, Motel, Gannery, Jacob and Dumas, themselves became important figures in the development of the chronometer, though they were dominated by the giant figure of Breguet, the great watch-maker who also worked on marine chronometers. By 1820 the basis of the perfect chrono-

38 Although this is a twentieth-century chronometer, the design has varied very little since the Harrison chonometer of 1735.

meter had been laid down, and the main task of later makers was in simplification and in ease of manufacture. In 1800 few vessels carried chronometers; by 1850 most well-equipped ships carried and used them.

However, the controversy between the chronometer makers and those who thought that timepieces would never replace bearings taken from the skies died hard, and until the end of the nineteenth century lunar distances were still observed, and tables of these distances did not disappear from the *Nautical Almanac* until well into the present century. The introduction of the electric telegraph helped navigation enormously, and time signals became more precise.

The most important advance in chronometer technology was in compensating for changes of temperature etc. The Royal Observatory tested a standard chronometer in 1886; at 50° F it had a weekly loss of 1·6 sec; at 92° F the loss was only 2·9 sec. Advances in astronomy and the establishment of the Admiralty Hydrographic Office in 1795 meant that more accurate and sophisticated navigational tables were published, and in 1832 tide tables were issued officially for the first time. As the nineteenth century drew on the speeds of the vessels were more accurately ascertained.

Save for decorative work on brass surfaces and such, navigational instruments are strictly functional, and because of the

39   *Fitness for purpose characterises navigational instruments, as in this double sextant.*

extremes of temperature and humidity that can exist on board ship where there is a choice between metal and other materials obviously metal is chosen. Marine instruments, though naturally reflecting design trends, are by their nature somewhat austere, and nineteenth-century instruments are completely lacking in the fripperies that bedevilled furniture and applied art. Even when there is scope for meaningless embellishment, as in compass cards, this has been rigorously eschewed. When one considers that it was not until 1760 that the chronometer definitively arrived, it is not surprising that eighteenth-century chronometers are expensive to collect. Chronometers come in the form of watches or clocks; the Victorian chronometer is plain, and usually set in a wooden case, with the controls set outside the rim of the clock.

Like the earliest examples of mathematical instruments, navigational instruments were usually imported into the United States in the eighteenth century, ordered from London instrument makers or taken there by settlers. Nevertheless, so great was the demand for angle-measuring instruments — triangulation was known in the 1750s in the United States — that many craftsmen in other fields turned their attention to the making of these. Surveying instruments were more acutely needed than navigational instruments in the early stages, though the threat of the Royal Navy speeded up the development of the United States Navy.

Many of the indigenous navigational instruments were made by clock-maker settlers, who because of their training were adept in the handling of metal and in the understanding of European instruments, which they copied. The shipping centres of America such as Boston were naturally the chosen homes of such makers involved in navigational instruments. One of the earliest immigrant instrument makers to arrive in Boston was John Dabney in 1739, who had been apprenticed to the instrument maker Sisson in London, and soon after Dabney was Anthony Lamb, who settled in New York about 1749 and advertised a 'newly invented quadrant for taking the latitude or other altitudes at sea'. Like Dabney, Lamb had been apprenticed to a London instrument maker, in this case Henry Carter.

Although it did not affect his skills as an instrument maker, Lamb had been an associate of the highwayman Jack Sheppard and had been sentenced to the gallows in 1724, a sentence that was commuted to transportation.

Philadelphia was also favoured by immigrant instrument makers. John Gould settled there in 1794. Robert Clark had 'received instruction under the best masters' in London, and made his home in Charleston where he made nautical and other instruments presumably not worried by competition (there does not appear to have been any other maker of navigational instruments in South Carolina until the nineteenth century).

One of the most important of navigational instrument makers in eighteenth-century America was Benjamin King of Salem, Massachusetts. His father had also made navigational instruments. The King family and William Hagger, of Newport, Rhode Island, specialised in sextants. Because of the absence of glassworks, and therefore the difficulty of obtaining optical glass, American instrument making suffered. There was also an absence of good mirrors, and it is not surprising that American sextants of the period are simple non-reflectors. Even in Britain, Hadley's octant was handicapped by a lack of good mirrors.

American instrument makers showed a good deal of ingenuity, and M. Morris of New York invented a 'nautical protractor' for a dollar, but this appears to have been merely a drawing-office instrument. In 1730 the Philadelphia maker Thomas Godfrey invented a species of back staff which was used in the survey of Delaware Bay. Not only was this improved back staff copied by other makers, but news of it got to the Royal Society in London and Godfrey received a prize in the shape of mixed household furniture.

Not surprisingly American clock-makers fought shy of the chronometer, and although their navigational instruments perfectly served the purposes for which they were designed there was nothing that they made that could not have been duplicated by a score of obscure London makers.

Navigational instruments, and, indeed, almost anything connected with the sea, are very collectable, and there is still ample scope for collectors on a modest budget. There are a number of interesting instruments used in the reading of charts, including the station pointer, which consists of a circle, from the centre of which project three long arms equipped with adjusting screws. The circle is calibrated. Station pointers were used to solve what is known as the three point problem — to locate on a plan the position of the instrument station when only three points were visible. Of the three arms, one is fixed, and the other two moveable, so that the correct direction of the three points can be tabulated. Sounding machines and gauges also exist in some quantity, and the patent log of 1812, based on an elongated propeller, with which the ship's speed was calculated by the number of turns in the cord pulling the propellor, was an ingenious answer to a problem that had perplexed navigators for centuries.

An unusual nautical instrument is the freeboard hydrometer, used for correcting the draught of vessels loading or unloading in fresh or brackish water. A freeboard hydrometer is basically a salinometer (for measuring the percentage of salt in water), a water-sampler with a graduated scale. An extremely interest-

40 *A distance-finder.*

ing little gadget is the distance-finder, which is made up of a moveable prism plate, a small sighting telescope, and an engraved base plate. The plate is divided by lines radiating from a central point rather like part of a sun ray, and is graduated in cables. The distance-finder was used to ascertain the distance of an object and for keeping a ship on station with another.

Navigational instruments have the advantage that they were made for hard use and in a variety of climatic conditions, and have thus survived the years very well, though early navigational instruments made of wood should on no account be overlooked if one is lucky enough to come across them.

CHAPTER 3

DIALS

DIALLING, SOMETIMES called gnomonics, is a branch of applied mathematics that deals with the construction of sundials. The first mention of the sundial is contained in the Bible, the first recorded sundial dating from about 300BC. This divided each period of daylight into twelve equal parts, known as temporary hours; as daylight varied from winter to summer this early sundial was not very accurate, though it survived until about AD 900 in an unchanged form.

A breakthrough was effected by Abu'l Hassan at the beginning of the thirteenth century, and the Arabs became pre-eminent in gnomonics just as they were in navigation. Abu'l Hassan speculated how to trace dials on cylindrical, conical and other surfaces, and introduced equal or equinoctial hours, but the concept of the temporary hour continued to be the one most in favour, and it needed the coming of the clock to revive interest in accurate sundials.

During the seventeenth century much was written about sundials, and a book of 1612 devotes 800 pages to this one topic, but by the eighteenth century the whole subject had been gone through and dismissed. The sundial became a garden ornament or an auxiliary to the church clock.

There are five types of sundial: horizontal dials: vertical dials on a vertical plane facing north, south, east, or west:

vertical declining dials on a vertical plane not facing cardinal points : inclining dials on planes neither vertical nor horizontal, subdivided into reclining (leaning away from an observer) and proclining (leaning towards an observer), and equinoctial dials, with the plane at right angles to the axis of the earth.

The most collectable type of sundial is, of course, the portable type, the equivalent of a pocket watch, and, equipped with a magnetic compass for orientation to the north, they were fashionable with the aristocracy during the seventeenth century. Many were of diptych form (i.e. hinged in two), and some were even built into wheel-lock pistols (the gimmick is not a twentieth-century innovation). Perhaps the most beautiful of these early dials was the solar compendium by Ulrich Schniep of 1575, in diptych form, with a multitude of tables including one for finding the solar time at different latitudes and a chart for casting

41   A late eighteenth-century Augsburg ring dial, sold in 1971 for £2,300.

horoscopes. This was sold in 1968 at Christie's for 4,800 guineas.

The ordinary diptych dials had one half connected to the other by a cord. When opened the 'lid' was at right-angles to the 'base', and the taut cord acted as the upright providing the shadow. This upright is known as the gnomon or the style. Diptych dials were mostly made in Germany, and especially at Augsburg and Nuremberg. Ivory and gilt metal were the materials most often used.

Perhaps the most collected dials of this period are those made by Michael Butterfield, who died in 1724. There is some doubt about the identity of Butterfield, and about whether he was a

42   *A selection of Butterfield dials.*

master instrument maker of German origin who settled in Paris and became instrument maker to Louis XIV, or an Englishman who had settled in Paris in early life. Simple as they are, there is no question why Butterfield dials are so much in demand, for elegance and function are splendidly combined. Collectors should be aware, however, that Butterfield dials are being faked in considerable numbers. The most sought-after dials are those with an octagonal base plate, made of silver, with the gnomon in the figure of a bird. The gnomon is adjustable, and can be used in several latitudes. The simplest forms of dials, without a graduated gnomon, were inaccurate more than a few miles from the place where they were designed for use.

Another of Butterfield's which is widely faked is his ring dial. A ring dial, also known as a poke dial, consisted of a ring with a sliding collar in which there was a small hole; the collar was adjusted to the solar declination and the ring suspended and positioned towards the sun, the light of which passed through this aperture and indicated the time on a scale engraved within the ring.

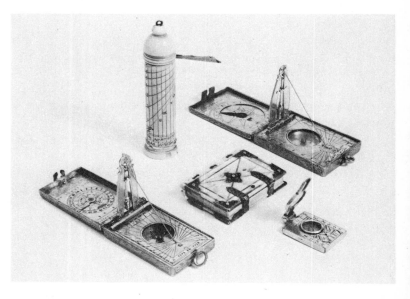

43 Diptych dials, a pillar dial, a pocket sundial, and a tablet dial in the form of a book dating from about 1610.

The Butterfield dial consisted of an engraved base plate, an adjustable gnomon, and a compass. Within the obvious limits of a hand-held instrument it was tolerably accurate. The only way for a dial without a compass to be accurate was to combine an ordinary horizontal dial with what is known as an analemmatic dial. The analemmatic dial has a gnomon that is moveable, and adjustable with the help of a scale. The two square dials are joined together, the XII and VI of both the dials in line. The combined dials are turned until they each show an identical time reading on their respective dial plates, and with the analemmatic gnomon adjusted the time is shown without reference to a compass.

Some of these analemmatic/horizontal dials are provided with screws for levelling purposes and a plumb bob, and English instrument makers were involved in their manufacture. Among them was Thomas Tuttle, sometimes spelled Tuttell, of Charing Cross, who flourished about 1700 and was responsible for high-quality ivory navigational instruments such as back staffs and quadrants. Despite the work of London and Paris makers, the centre of the portable-dial industry lay in Germany with such practitioners as Schissler, Johann Martin (1642–1720), Johann Willebrand (died 1726), Nicholas Rugendas, and Tobias Volckmer of Munich.

Martin and Willebrand were prominent in the making of crescent dials, self-orientating pocket dials of silver and gilt-brass, which derived from the universal equinoctial ring dial. An equinoctial dial is one in which the hour scale is set parallel to the plane of the equator. In the crescent dial the hour ring is cut into two semicircles set straight edge to straight edge, and the instrument gets its name from the crescent-shaped gnomon which slides on a scale.

British instrument makers were more prominent in larger universal equinoctial (also known as equatorial) dials; there exist fine examples of equinoctials by George Adams, instrument maker to King George III, and they are elaborate precision-made instruments. Some of these have mechanical working parts, and William Deane of Crane Court, Fleet Street, made a particularly fine one about 1690, complete with the most lavish baroque ornamentation.

In appearance equinoctials have more in common with armillary spheres (pp 102–3) than with the ordinary sundial. They are essentially three dimensional. Unquestionably they were made to serve a purpose, but this purpose was subsidiary to

44 *Equinoctials and armillary spheres were library pieces rather than viable scientific instruments. This nineteenth-century armillary sphere was surely undervalued, however, when it only made £150 at auction.*

their being prestige instruments for rich amateurs, library pieces rather than outdoor time-tellers. In the wider sense, sundials have always appealed to amateurs, both in the collection and the making, and the problems of inscribing graduations upon cylinders or cubes have intrigued many right up to the present day, including the sculptor Henry Moore who created a sundial of an abstract equinoctial shape for the forecourt of *The Times* buildings in 1967. It is interesting that although sundials were logically obsolete upon the introduction of an accurate clock, the French railways continued to use a sundial (defensively called a heliochronometer) until 1901 against which railway clocks were set.

Spherical dials in which the sphere was a replica of the earth showed what parts of the earth were bathed in sunshine, but could hardly be reckoned a viable time-teller, and dials in the forms of crosses and stars, interesting as they no doubt are, were unnecessarily complex ways of doing a job an ordinary garden sundial could do much better. Pillar dials, in which the graduations ran in ellipses around the circumference of a cylinder, might be termed philosophical curios rather than functional objects.

The nineteenth century renamed sundials solar chronometers, and a number of interesting models were made incorporating not only an analemmatic device but a Vernier scale. In terms of pure instrument making perhaps the finest was that devised by Sir Charles Wheatstone, co-inventor of the electric telegraph, and made by Elliott of London, where the time is indicated on a dial in hours and minutes as in a watch. Via a lens set at the end of a tube, an image of the sun fell on the centre of a system of concentric circles. Two Vernier scales were incorporated, one for setting the polar axis for latitude and the other for adjusting the tube for the sun's declination. Essentially the Wheatstone solar chronometer was a deliberate venture into a scientific backwater.

Surprisingly, sundials have been made until quite recently to tell the time in places where watches and clocks are in short supply. The peasants in the Pyrenees made pillar dials, and sometimes pillar dials go by the name of shepherds' dials. Polyhedral dials, usually in the form of cubes, were popular in the eighteenth century, and can occasionally be seen in old-world gardens.

CHAPTER 4

TELESCOPES

THERE ARE two kinds of telescope, the refractor and the reflector, and they work on two different principles. The refracting telescope was discovered by accident in 1608 by a Dutch spectacle maker. Some accounts say that two children were playing in his shop and noticed that by holding two lenses in a certain position the weather-vane of a nearby church appeared nearer. The Dutch spectacle maker, a man named Lippershey, wasted no time in mounting the lenses in a tube, and so the telescope was born, to this day unaltered and unalterable.

There is some doubt whether he used two convex lenses, or a convex and a concave. If he had used two convex lenses he would have seen the steeple upside down. Dismissed by a contemporary as an illiterate mechanic, Lippershey showed sufficient acumen to approach the government, which welcomed the telescope as an aid to warfare. The news soon spread throughout Europe, and reached Galileo in 1609, who designed what is now known as the Galilean telescope, consisting of a plano-convex object glass (the lens at the far end of the tube) and a plano-concave eyepiece mounted in a lead tube.

The discovery of the telescope was inevitable once the properties of curved glass had been appreciated. Roger Bacon (1210–94) had studied the effects of mirrors and plano-convex lenses, and seems to have used a hand magnifier as an aid to vision, but the invention of spectacles is credited to Salvino d'Armati (died 1317) on the evidence of the inscription on his tombstone in Florence — 'Inventor of Spectacles. God pardon his sins.'

Not surprisingly Galileo was more interested in using his telescope to study the heavens than for military purposes, and although he discovered four satellites of the planet Jupiter and viewed the surface of the Moon, as well as observing sunspots, the quality of the glass was inferior and he was troubled with the two problems that continued to dog telescope makers — spherical aberration and chromatic aberration. Spherical aberration arises, naturally enough, from the use of spherical surfaces for the lenses. Rays on the edge of the lens are more refracted than those nearer the centre. Chromatic aberration arises from a phenomenon known as dispersion; light of the various colours of the spectrum are refracted by different amounts on passing through the lens.

Until Newton found that white light was composed of the colours of the spectrum, chromatic aberration was a mystery, and astronomers and others using telescopes had to make do as best as they could. They found that by reducing the curvature of the object glass (thus increasing the focal length) they could reduce the depressing effects of both spherical and chromatic aberration. But this brought its own chain of problems, for telescopes became longer and longer, culminating in telescopes 150ft long, or in the bizarre aerial telescope of Christian Huygens in which, in effect, an immensely long telescope had its centre taken out, leaving an eye piece and an object lens set high in the sky, connected with the eye piece only by a length of silk.

By 1610 Galileo had made a telescope that magnified 30 times; it was yet to be named a telescope, and Galileo called his invention a perspicillum. The first printed record of the word telescope appeared in 1612, but by this time Galileo was not the only one in the field, and there is a good deal to suggest that Thomas Harriot, mathematical tutor to Sir Walter Raleigh,

45   Astronomy and the study of the heavens was a leisure activity for gentlemen, well illustrated in this picture by Wright of Derby.

used what was described as 'a perspective glass' as early as 1610 that was identical with the Galilean telescope. In 1638 Galileo went blind, and imparted his method of grinding and polishing lenses. It was found that he used only the centre portions of his object glasses, improving the sharpness of the image but losing light.

The great breakthrough in refractor telescopes was made by Christian Huygens (1629–93), for not only was he the developer of the aerial telescope but, more important to the development of the telescope, he cut down chromatic aberration by replacing the single eyepiece with two thin convex lenses. Without realising it he had stumbled on the formula for the present day eyepiece, though today the eyepiece consists of two plano-convex lenses of crown glass, one with a focal length of two to three times that of the other and separated by a distance equal to half the sum of their focal lengths.

Huygens was not the first to use more than one lens for the eyepiece. His predecessors used five, eight, or even nineteen spaced convex lenses. None of these optical systems was effective. Huygens put a lot of time, research, and energy into the business of grinding and polishing lenses, and communicated his findings to the Royal Society, though the more critical of the fellows looked askance at Huygens' method of smoking his eyepiece to get rid of chromatic aberration.

The largest object glasses were made in Italy, and in the last half of the seventeenth century telescopes became more and more powerful and cumbersome. It was a Frenchman, however, who thought up a glass with a potential of 1,000 magnification; the aim of this telescope was to see animals on the moon, but it was completely out of reach of the seventeenth century, and there was a limit to the size of telescopes, notwithstanding how much ingenuity and money went into the mounting of them.

The unmanageability of long telescopes caused inventors to consider the idea of a fixed telescope with the image fed to the object glass by means of a travelling mirror, but if good lenses were difficult to make so were good mirrors. The heliostat was thought up in 1682 and a passable model was made in 1720, but it was never very successful. The desirability of not only picking up stars but following them round the heavens also spurred on inventors into mounting the object glass upon a clockwork drive and following the image by moving the eyepiece support, an innovation of 1685 that did not catch on.

The large telescopes of the seventeenth century were clearly of little practical use except in astronomy, and there was a much greater demand for 'perspective glasses', as the simple unsophisticated Galilean telescope was still called in England. Seventeenth-century dilettantes were very science conscious, but London opticians and instrument makers had neither the skill nor the glass, and perspective glasses, also known as prospect glasses, were imported from the continent. But eventually the English makers demonstrated that they could vie with the Italians and the Dutch and before the start of the eighteenth

46   *A simple Galilean telescope on a mount.*

century many firms and makers were producing telescopes; these included John Marshall—the most important—Christopher Cock, John Cox, Joseph Howe, James Mann, Richard Reeves (or Reive), Edward Scarlett and John Yarwell.

Telescopes of this period are still in existence and naturally, when they turn up at auction, they command very high prices. Those up to 3ft focus have one tube, but longer telescopes are made in sections (telescopic). The tubes are usually of card and covered with shagreen, silk, leather, or parchment, with wood and ivory fittings. Because of their length they are kept light in weight.

A new avenue of approach was opened in 1666. The year of the great fire of London was also the year when Newton passed sunlight through a glass prism, and having found that white light broke into colours he was a step forward to finding out the exact nature of chromatic and spherical aberrations. It seemed to him that conventional telescopes would always be defective by virtue of these aberrations. Lenses could not be made free of them; but a system of mirrors could be, or reasonably so.

In 1663 James Gregory had also postulated the reflecting telescope. A large paraboloidal concave (a) would collect the

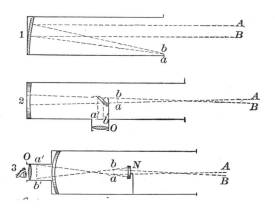

47 *The various forms of reflecting telescopes.*

light and reflect it to a small ellipsoidal concave (b), which would form the image at the centre of (a), perforated for the insertion of an eyepiece. James Gregory employed Richard Reeves (or Reive) and John Cox to create a suitable mirror, but they failed to do so and Gregory went off in disgust, giving his name to the Gregorian telescope, and leaving Newton with a clear field.

Like Gregory, Newton used a concave paraboloid as his main mirror, but instead of a secondary ellipsoid he used a small flat mirror which reflected the image to an eyepiece in the side of the tube. This first reflecting telescope was no more efficient than the first refracting telescopes.

As collectors of antiques will know, mirrors of this period are not very good, and a superb reflecting surface was hard to come by. Newton chose an alloy known as bell metal—six

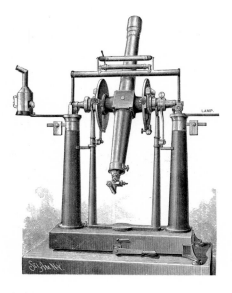

48 A transit instrument.

parts of copper to two of tin, plus one part of arsenic. The high copper content caused the mirrors to tarnish rapidly, and they had to be kept continually polished.

Another method of making a reflecting telescope was being tried out in France, the invention of Cassegrain, and it was suggested that this antedated the Newtonian reflector and was better. It was more akin to Gregory's telescope than Newton's. In place of Gregory's secondary concave mirror, the Frenchman used a secondary convex mirror. The image arrived at an eyepiece in the centre of the main concave mirror very much as in the Gregorian telescope, but notwithstanding the prestige involved, Cassegrain retired from the lists.

The Newtonian telescope created a furore among the scientists of Europe, but no one was in any doubt about the poor quality of the reflecting surfaces. The Newtonian mirrors lost more than 80 per cent of the incident light, and although Robert Hooke did a good deal of experimentation with mirrors both of glass and metal, it seemed as though the Newtonian telescope would remain in a primitive state through the opticians' inability to make adequate mirrors, and that the refracting telescope, despite built-in disadvantages, would win the day. It was certain that for mariners and soldiers there was no point in further experimentation, and only with the introduction of prismatic binoculars did their adherence to the Galilean system falter.

The eighteenth century opened with the refracting telescope supreme. Great ingenuity was shown in reducing the size of telescopes by telescopic tubes, and a 5ft telescope could extend to more than 16ft. Telescopes were frequently supplied with two object glasses, one for use in the day, one for use at night. Most telescopes now had double or triple lens eyepieces, and for night work one of the lenses was taken out.

About 1719 the young scientist John Hadley (inventor of the reflecting octant) made a telescope on the Newtonian model, presented it to the Royal Society in 1721, then in 1726 made a Gregorian telescope. It was matched with one of Huygen's monster telescopes, and although the Huygen's gave a brighter image, the definition of Hadley's instrument was as clear. With the renewal of interest in reflecting telescopes, work was begun again on the vexed question of a good mirror, and more than 150 different alloys were tried. By the middle of the eighteenth century many opticians were making good quality reflecting telescopes, including Scarlett, Hearne, James Mann, Chaplain and James Short.

The best of these were by James Short, who anticipated later technology by using glass backed with quicksilver, and despite Short living and working in Scotland his products were compared with those of his contemporaries working in London, usually to their disadvantage. London was becoming the centre of the instrument-making world, and soon Short came to London, with a workshop in Surrey Street, off the Strand. Most of his telescopes were Gregorians, varying in size from small hand-held models to very large ones with an 18in aperture and 12ft focus. To the delight of those who can afford to buy his work, Short had the welcome habit of not only inscribing his telescopes with his name but the serial number; each of his telescopes has the equivalent of an opus number. The highest serial number known is 1370. He had a more regrettable habit, so far as his customers were concerned, of over-rating the powers of his instruments, and an observatory reflector with a claimed power of 800 was found to have a power of only 130.

The reason for excessive claims by Short was that the greater the magnification the higher the price. A 130 magnification reflector cost 20 guineas, 800 magnification cost 300 guineas. Perhaps the largest instrument he constructed was that made for the King of Spain in 1752; with an 18in aperture it cost £1,200.

49 *By the middle of the eighteenth century brass was becoming the main material of the telescope makers, though shagreen was still used on mounted telescopes.*

Short's telescopes were of brass, and from the middle of the eighteenth century this became the predominant material of the instrument makers, due to improved methods of processing brass and better tool-making technology. Although Short's telescopes were not so fine as they might have been—it is certain that he overproduced—they were a great advance on the cumbersome and erratic instruments of a previous age. They were nearly all mounted on altazimuth stands, which meant that they could be manoeuvred through both vertical and horizontal axes, and he invented a 'universal' portable mount for small telescopes. But he earned his success through his fine mirrors, the secret of which he never disclosed, though some of this success with mirrors was the product of common sense. He made a good many mirrors of the same focus, and tried them one after the other till he found a good match.

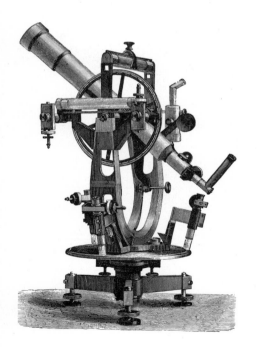

50 *An altitude and azimuth instrument—virtually a large theodolite.*

TELESCOPES

FRANCIS MORGAN,
Optical, Philosophical, and Mathematical Instrument-Maker,

At the Sign of *Archimedes* and *Three Spectacles*,
No. 27, *Ludgate-Street*, near *St. Paul's*,
LONDON.

MAKES and Sells Spectacles of Glass or Pebble, in neat and light Frames of the newest and best Construction, and in the most convenient Manner for avoiding Pressure on the Nose or Temples.

Concaves for very short-sighted Persons; and Glasses for Reading, Magnifying, and Burning.

Reflecting Telescopes, of all Kinds, and with the latest Improvements.

Refracting Telescopes of all sorts; with one of a New Invention (useful for Sea or Land) that will not warp in any Weather; and a Sort to use at Sea in the Night.

Double, Single, Solar, Opake, or Aquatic Microscopes.

Camera Obscuras, to delineate Landscapes and Prospects (and which serve to view Perspective Prints) made truly parallel; Sky-Optic Balls; Prisms, to demonstrate the Theory of Light and Colours; Concave, Convex, and Cylindrical Speculums; Magical Lanthorns; Opera Glasses; Optical Machines for perspective Prints; Cylinders and Cylindrical Pictures.

Air-Pumps and Air-Fountains of various Kinds; Glass taps; Portable Apparatus for Electrical Experiments, which is allow'd by the Curious to be the best of the Kind.

Barometers, Diagonal, Standard, or Portable.

Thermometers properly and accurately adjusted.

Hygrometers; Hydrostatical Balances, and Hydrometers.

Hadley's Quadrant, after the most exact Method, with Glasses truly parallel; Davis's Quadrant. Globes of all Sizes; Compasses, Azimuths, Steering, and Plain; Load-Stones; Nocturnal and Sun Dials of all Sorts.

Scales; Cases of Drawing-Instruments; Parallel Rulers; Proportional Compasses; and Drawing Pens.

Theodolites, Semicircles, Circumferenters, Measuring-Wheels, Spirit Levels, Rules, and all Sorts of the best Black-lead Pencils, and all other Sorts of Instruments of the newest and most approved Invention.

FRANÇOIS MORGAN,
OPTICIEN,

A l'Enseigne de l'*Archimède* et *Trois Lunettes*, au Numero 27, dans la Ruë de *Ludgate*, proche de l'Eglise de *St. Paul*, à LONDRES.

FAIT et Vend des Lunettes de toutes les differentes Façons, proprement et commodement montées pour éviter toute Pressure sur le Nez ou les Temples.

Verres à lire, à grossir considerablement les Objets, et à bruler.

Telescopes à Mirroir de toutes Sortes, dans la derniere Perfection.

Telescopes de Refraction en toute genre, particulierement d'une nouvelle Construction (qui servent également par Terre et par Mer) qui resistent à l'humidité sans se voiler, il en fait aussi pour la Nuit en Mer.

Microscopes doubles, simples, solaires, opaques, ou aquatiques.

Chambres Obscures, propres à dessiner Païsages et Perspectives, ou à tirer des Plans (qui servent aussi pour à voir des Estampes de Païsages, et de Perspectives) les Mirroirs desquels ont les surfaces parfaitement paralelles et plates; des Yeux de Beufs; Boules Optiques; Prismes pour demontrer la Theorie de la Lumiere et des Couleurs; Mirroirs concaves, convexes, et cylindriques; Lanternes Magiques; Lunettes d'Opera; Machines Optiques pour Estampes de Perspectives; Cylindres, et Tableaux Cylindriques.

Pompes Pneumatiques; Fontaines et Jets d'Eau, qui jouent par l'Air, de diverses Sortes, avec une Machine Electrique portable, que les connoisseurs reconnoissent pour la plus parfaite et la plus commode qui ait été faite jusqu'à present.

Barometres diagonaux, portables, ou fixés.

Thermometres, dont les Degrés sont proportionés à l'ouverture des Tubes.

Hygrometres, Balances hydrostatiques, et Hydrometres de toutes Façons.

Octants de Monf. Hadley, avec des Mirroirs exactement paralele et plates; Quadrans de Davis; Globes Terrestres et Celestes de toutes Grandeurs; Boussoles d'Azimuth de toutes Especes; Aimants artificiels; Nocturlabes; Cadrans Solaires de toutes Sortes.

Etuis d'Instrumens mathematiques et pour dessiner; Regles exactement divisées, et à Lignes paraleles; Compasses de Proportion, et Plumes à dessiner.

Theodolites, Demi Cercles, Circonferentier, Roués à mesurer Distances, Niveaux, Regles, Crayons de Mine de Plomb des meilleurs, et toutes autres Sortes d'Instrumens de la plus nouvelle et plus approuvée Construction.

CURIOUS OLD HANDBILL IN ENGLISH AND FRENCH.

51   By the eighteenth century London instrument makers had acquired great prestige, and Francis Morgan issued his handbills in both French and English. Notice Archimedes with a telescope.

Short, who died in 1768, had corresponded with John Mudge, a doctor interested in mirror making. Mudge researched deeply into the best alloy for metal mirrors, and in 1777 was awarded a medal by the Royal Society. He used the same basics as Newton—copper and tin—but in slightly different proportions and without additives, but he realised the key to the perfect mirror was in grinding and polishing which he brought to a fine art, communicating his findings to the instrument maker Joseph Jackson, who had workshops in Angel Court, Strand. Mudge was not the only man investigating alloys, and iron, lead, bismuth, brass and zinc were all tested, but even the brightest metal mirror became rapidly tarnished on exposure to the atmosphere.

By the middle of the eighteenth century, instrument making, a craft located mainly in or off the Strand, was well established. Makers were kept busy on microscopes, telescopes, and on surveying instruments demanded by the increase in trade and commerce, and therefore sea communication.

William Herschel was outside this circle. Born in 1738, he was an amateur mathematician and astronomer. He found that the instruments on sale did not satisfy him, and in 1774 constructed a large Gregorian reflector, followed in 1775 by a Newtonian reflector. His fame was established in 1781 when he discovered Uranus, and he was made Royal Astronomer the

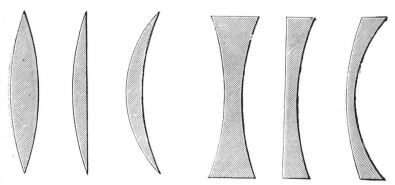

52 *The various types of lenses used in telescopes and microscopes.*

following year, but despite his elevation he continued to grind and polish his own lenses and even make and sell astronomical telescopes. He made and sold about 60 reflecting telescopes. In 1789 he made the largest telescope in the world, a 40ft reflector with a 48in aperture, receiving £4,000 from the King, plus £200 a year for its upkeep. It was something of a white elephant, and Herschel preferred his smaller pieces, accepting that the 40ft telescope was a royal status symbol.

As has been mentioned, the eighteenth century opened with the refracting telescope supreme, a supremacy not disturbed until 1726 when Hadley made his Gregorian reflector, and it seemed as though technological advances in the reflector would nullify the inherent advantages of the refractor, i.e. utility and portability. In 1729 Chester Moor Hall combined two glasses of opposite powers, a concave lens of flint and a convex of crown glass. The achromatic ('Free from colour; not showing colour from decomposition in transmitting light'—OED) lens was born, though the name was not applied to it until 1766.

Hall does not appear to have ground his own lenses, but used Edward Scarlett of Soho and James Mann senior of Ludgate Street. These two opticians subcontracted to George Bass of Bridewell Precinct. The flint and the crown glass lenses arrived independently, but Bass put them together and discovered the new breakthrough in optical technology. Unlike Short in the field of the reflector, Hall was not secretive, and John Bird of the Strand and Ayscough heard of the achromatic lens, though without appreciating the potential, and it was only when John Dollond heard from George Bass that the achromatic lens began to receive its due. In 1752 John Dollond and his son Peter set up shop at the Sign of the Golden Spectacles and Sea Quadrant near Exeter Change in the Strand, and applied for a patent for the new telescope, covering

> the new invented method of refracting telescopes, by corresponding mediums of different refractive qualities, whereby the errors arising from the different refrangibility of light,

53 *The trade card of Henry Pyefinch, one of the makers sued by Dollond for infringing his patent.*

as well as those which are produced by the spherical surfaces of the glasses, are perfectly corrected.

In 1761 John Dollond died, and a partner that the Dollonds had taken into the firm left, and, in defiance of the patent, made and sold his own achromatic telescopes, to the distress of the other London instrument makers. To add insult to injury, they were warned that if they sold achromatic telescopes without paying him royalty they would be sued. The opticians clubbed together and petitioned the Privy Council to revoke the patent. This petition was unsuccessful, and many opticians did make achromatic telescopes and were sued by Peter Dollond and his erstwhile partner; these makers included James Champneys of Cornhill, Francis Watkins and Addison Smith of St Martin's Lane, and Henry Pyefinch of Cornhill.

The Dollond patent became void in 1772, but opticians by and large were slow to take advantage of this. There was difficulty in obtaining blanks of flint glass free from flaws and there was a high import duty on such blanks. In the meantime Peter Dollond was experimenting in triple object glasses to get rid of a secondary spectrum that had arisen from the two lenses of flint and crown glass. His first triple objective dates from about 1763.

The Dollonds are two of the most important figures in instrument making. John Dollond was the theoretician, Peter Dollond was an exponent of trial and error, and it is interesting that their telescopes still turn up at reasonable prices. One brass hand-held Dollond telescope changed hands in South Devon in 1972 for £6.

In 1766 Peter Dollond moved to new premises at 59 St Paul's Churchyard, where he was joined by his brother John. In 1783 they began to use brass draw-tubes instead of paper-covered vellum, though the patent for these was taken out in May 1782 by Joshua Martin. Compared with earlier telescopes the Dollond instruments were very cheap; small telescopes cost as little as £2.

54 Martin, the maker of this instrument, took out a patent for brass drawtubes in 1782 that was widely disregarded.

The Napoleonic wars brought a huge demand from the army and the navy, and Dollond introduced his 'Army telescope' in various sizes, 14 to 52in, priced between $2\frac{1}{2}$ and 12 guineas. All scientific instruments enjoyed a boom, and Dollond made sextants and theodolites, as well as more mundane things such as spectacles. Peter Dollond died in 1820, but the business was kept going by his nephew George Huggins, who, unfortunately for collectors, called himself George Dollond.

More versatile than the Dollonds, and historically more significant, was Jesse Ramsden, whom we have met in the chapter on surveying instruments. Born in 1735, Ramsden married a daughter of John Dollond, and therefore it is not surprising that their products are linked. Apprenticed to Burton, an instrument maker in Denmark Street, Ramsden set up in the Haymarket in 1762, and was involved in the mountings of Short's reflecting telescopes and Dollond's instruments before entering the field of reflecting telescopes, using the Cassegrain method.

TELESCOPES

The great thing about Ramsden's telescopes is that they were utterly stable, and his work was in great demand by observatories; at one time he had fifty workmen in his employ, and made barometers, pyrometers, surveying chains, balances, levels, as well as ordinary drawing instruments. He died in 1800, leaving an immense gap in the scientific spectrum.

Among his contemporaries were not only Dollond, but Edward Nairne, born in 1726 and operating from 20 Cornhill, maker of barometers, air-pumps, levels and thermometers, Thomas Blunt, once an apprentice of Nairne, who opened workshops at 22 Cornhill, and William Cary, who set up business in the Strand, who made fine instruments for surveying and astronomy. There was also Thomas Jones, a former employee of Ramsden, who had workshops at 62 Charing Cross.

More important than these, however, were the Troughton brothers, Edward and John, and William Simms. Edward Troughton was born in 1753, and in 1770 joined his elder brother John, who was then managing an instrument-making

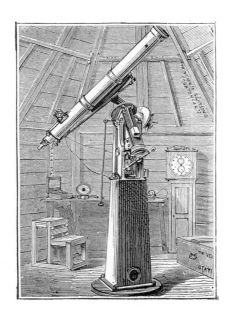

55 *An equatorial telescope used in Cairo for the transit of Venus in 1874.*

TELESCOPES

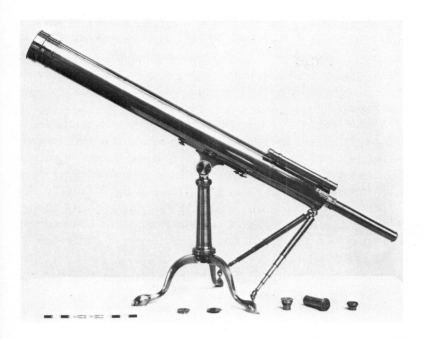

56 Thomas Jones was a former employee of Jesse Ramsden, and produced some fine instruments, such as this refractor.

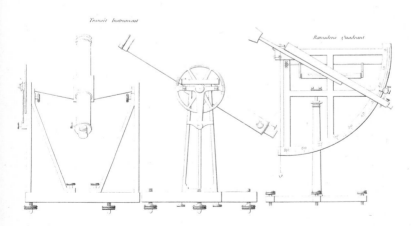

57 Diagrams of a transit instrument and Ramsden's quadrant.

business in Surrey Street, off the Strand. The brothers set up trade at 136 Fleet Street in 1782, and they made a wide variety of instruments including transit circles (telescopes fixed in the centre of an axis, supported at both ends by pillars, and fitted with large circles). They were purely tracking instruments and the telescopes were of no great power. The Troughtons also made numbers of surveying and navigating instruments.

Edward Troughton, the most important of the brothers, was something of a character. A man of frugal habits, he never married, and spent most of his life in his back parlour wearing snuff-stained clothes and wig, ear trumpet in hand. Late in life, in 1826, he took William Simms into partnership, and it was he who, under the mark Troughton and Simms, made the important 36in theodolite for the great survey of India. At the time it was considered the best theodolite ever made. After Troughton had gone, Simms took his nephew into the firm, and the name was kept alive.

By the end of the eighteenth century, instrument making had reached its zenith, but although the mechanical side and the metalwork were superb, the opticians were still handicapped by poor glass. English glassmakers were lucky if they made discs larger than four inches in diameter, and as they reserved their best efforts for the prestige makers, it was difficult for newcomers to make headway. Also there was a penalising excise duty on flint and crown glass, unfortunate as the Europeans were making great progress in producing large discs free from faults, and Charles Tulley, who became the major figure in telescope making after the death of Peter Dollond, used French glass for some of his better instruments. Tulley started business in Islington about 1782, and died in 1830, whereupon his sons took over the business. The Tulley business seems to have died out about 1843.

Perhaps the finest lens Charles Tulley made was for the first telescope in Britain to be driven by clockwork, initially erected in 1829 and used by the British Government to observe the transit of Venus in 1874. The flint blank for the lens was obtained from Paris in 1828, the construction was by George Dollond, alias Huggins, who also made the eyepieces which afforded a range of magnification from 22 to 1,200.

Despite the disadvantages of not having glassworks, the Americans succeeded in making many excellent telescopes before the close of the eighteenth century, such as the transit telescope by David Rittenhouse used to view the transit of Venus in 1769. This was quite a small instrument with an aperture of less than 2in and a focal length of 32in. The instrument maker Henry Voigt also made a telescope of high quality in 1790. However, reflecting telescopes had to wait until the nineteenth century before they appeared in the United States. Whereas it was possible for clock makers, and even cabinet makers and other skilled craftsmen, to turn their talents to some categories of instrument making when they emigrated to the United States, these categories excluded most types of instrument where optics were involved. It was a field where amateurs ventured at their peril.

In the early years of the nineteenth century there were a number of British amateurs making and selling reflecting telescopes, among them James Veitch of Inchbonny in Scotland. John Ramage, an Aberdeen tradesman, had the backing of the Astronomical Society in his attempt to make huge astronomical reflectors. It failed, and in 1839 the largest telescope in the country was Herschel's 1789 reflector, now in a sad and dangerous state.

This state of affairs did not last for long, and the early years of Queen Victoria's reign saw an upsurge in giant telescopes, due to improved methods of making mirrors. James Nasmyth, famous as the inventor of the steam hammer, made in 1842 a very large instrument on an altazimuth mount. The eyepiece arrangement enabled the user to remain seated without moving his eye as the body of the telescope was manoeuvred. The massive tube was made of sheet iron. Even more spectacular was the Rosse telescope, completed in 1845, with a mirror 6ft in diameter. Until a 100in telescope was located at Mount

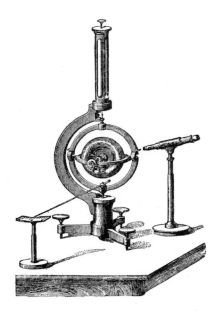

58 *A gyroscope, used for demonstrating rotation of the earth.*

Wilson in California in 1919 it was the largest telescope in the world. Like the Herschel telescope, weather conditions meant that for most of the year it was unusable, and its great size made it useless as a tracking instrument.

Another maker of large telescopes of the period was William Lassell, and he produced fine mirrors by getting the perfect proportion of copper and tin. In 1852 Lassell took a large telescope to Malta where climatic conditions were better. Warren de la Rue was another telescope maker, who created astonishment by his photography of the heavens from 1852 onwards.

Many of the instruments produced in the mid-nineteenth century proved to have an astonishing life span. Troughton had produced the first adequate transit telescope in 1806; by 1848 this had become erratic through continuous use, and George Airy designed another one, with an object glass by Simms (costing £300). The transit telescope was not a mere amateur's toy, as many of the astronomical telescopes of the nineteenth century were. It was set up with its axis in the plane of the meridian, and used to note the passage of stars and planets across the meridian. By concentrating on stars whose positions were known, the transit telescope provided a valuable time-check. Airy's transit instrument was used until 1922 at Greenwich, the instrumental basis of time-keeping in company with a chronometer.

Britain had enjoyed a supremacy in instrument making in the late eighteenth and early nineteenth century, and the industry's products were exported all over the world, America proving an especially profitable market. By 1850 the Americans were beginning to make their own instruments, the Germans were busy — the London instrument makers C. May and W. R. Dawes were importing Munich object glasses — and provincial instrument makers, such as Thomas Cooke of York, were beginning to impinge on the metropolitan market.

1855 was a key year for instrument making. Cooke started the first telescope factory at Bishopshill, Yorkshire. He found a new maker of optical glass, Chance Brothers of Birmingham. The Chance Brothers had benefited from the 1848 revolution in France, which caused many French glassmakers to come to Britain, where they were taken up by the Chance Brothers. Cooke produced a great many high-quality telescopes at a reasonable price, for he could undercut the prestige London makers by production-flow methods. He cast all his own brass and, a skilled engineer as well as a fine optician, installed his own machine tools.

Notwithstanding the publicity that was given to the massive reflecting telescopes of the mid-nineteenth century, the refractor was the chosen instrument of both official and amateur astronomers. Lenses were everywhere improving, due to the endeavours of people like the Chance Brothers utilising modern methods of production, but the metal mirrors did not seem susceptible to any great improvement. The 1851 Great Exhibition opened up possibilities. An acute observer at one of the stands noted the globes and vases silvered on the inside to a recipe patented by Messrs Varnish and Mellish, but the implications were seen by the French and Germans rather than the British, and they were in the van with their silvered-glass mirrors, rendering the metal mirrors obsolete overnight. It was the technical breakthrough that the advocates of the reflecting telescope had been waiting for for centuries.

In the early 1860s George With produced three or four silvered-glass mirrors of the highest quality which were examined and approved by the Royal Astronomical Society, and also involved in mirror-making was George Calver of Chelmsford, best known for a 36in mirror that was mounted in a Newtonian reflector in 1879. Using this telescope a photograph of the Orion nebula won the owner the Gold Medal of the Royal Astronomical Society.

59 Improvements in mirrors made reflecting telescopes popular amongst amateurs in the 1860s. This illustration shows a typical nineteenth-century instrument. It was sold recently for £140.

The introduction of the silvered-glass mirror into astronomical telescopes in 1856 shifted interest back into reflecting telescopes, and although the refracting telescope never lost its appeal for amateurs all serious astronomical work was thereafter accomplished using reflecting telescopes.

TELESCOPES

Telescopes, of course, are frequently used in combination with other devices such as theodolites or levels. For astronomical use they are often incorporated into an elaborate framework to carry out specific duties, such as tracking a planet or measuring the angles of various stars. The transit instrument consists of a telescope, moveable in the plane of the meridian upon a horizontal axis, for the purpose of observing the precise instant when any celestial body passes the meridian. The equatorial telescope is a tracking instrument, and the mounting of large equatorials takes into consideration the fact that while a star is held in focus the earth itself is moving; this movement is countered by a clockwork mechanism. Once a star has been located the equatorial will continue to follow it automatically. Altazimuth mountings enable the telescope to be moved vertically and horizontally, but normally would not permit the tracking of a star.

An astronomical quadrant, not to be confused with the navigational variety, is a pair of telescopes attached to a quadrant, and is not often met with nowadays. Its role was taken over by more versatile instruments. The advantage of a quarter-circle over a full circle in the early days was that a quarter-circle

60 Left. *A transit instrument by W. Jones.*
61 Right. *An equatorial telescope by Dollond.*

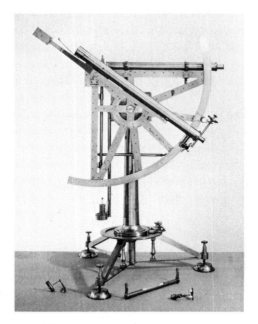

62   *A quadrant by Dollond.*

could be constructed on a large scale (one by the late-eighteenth-century maker Bird was eight feet across), and this meant that the graduated scale of 90° had wide divisions, facilitating direct readings.

In the days of Nelson and the Duke of Wellington the hand telescope was a valuable aid, but they were still cumbersome things on ships or on the battlefield, though gone were the days when telescopes had eyepieces larger than the objectives, and were so long that they had to be made from covered card to enable them to be lifted up, let alone used in tricky situations. The longer a telescope was, the more difficult it was to keep an object in the field of vision.

In 1823, the Vienna optical master Voigtländer, responsible for camera lenses that were used unchanged for half a century, conceived the idea of two telescopes fixed side by side, each telescope individually focussed. In 1825 J. P. Lemaire brought in the idea of a central focussing device. Nevertheless true field glasses had to await 1859 for the Frenchman A. A. Boulanger to introduce prismatic binoculars, and with the coming of these the naval and military telescope went into decline, though the telescope continued to be the mark of office of the officer of the watch. Magnification of the order of ×8 or ×10 was quite sufficient on the battlefield.

In 1893 Ernst Abbé disposed the prisms in field glasses in a different way, so that the object lenses were further apart while the advantages of binocular vision were preserved, and with this innovation binoculars came of age. Only in opera glasses were the combinations of object lenses and eyepieces, without the interpolation of prisms, preserved.

CHAPTER 5

SPHERES AND ORRERIES

TERRESTRIAL AND celestial globes have always been prestige items, and were the basic furniture of the gentleman's library irrespective of whether or not he knew one end of a telescope from the other or whether the tropics were fashionable diseases or not. To some extent, the kudos of having a globe about diminished during the nineteenth century when there was a big demand by schools for globes of all sorts and sizes, and by educational institutes such as the polytechnics, anxious to keep abreast with the plethora of astronomical discoveries. In the nineteenth century astronomy was an extremely popular subject, and there was no better way to instruct the unlettered than by means of a celestial sphere. It was really the only way, for star maps, being on the flat, are as misleading as Mercator's projection in geography.

Globes to represent the stars in the sky have probably been used for 2,500 years, and the earliest globe extant dates from about 200 BC. To most men living before Copernicus it seemed a self-evident fact that the stars were fixed to the surface of a revolving sphere, and although representing them on a globe meant that they were inside out this seemed the best way to demonstrate their position. Not until such buildings as the Baker Street Planetarium were conceived was the real answer to the problem of portraying the heavens found.

There are two ways of putting the stars on a globe; (a) by hand, in which case the globe was a one-off and correspondingly rare; (b) by printing. As is obvious, printing can only be done on a plane surface, but as early as 1500 gores were used. A gore is a piece of paper of a special shape, pointed at the top and bottom and wider in the middle. By sticking a dozen or so of these printed gores on a sphere the entire surface would be satisfactorily covered, and though, in the primitive state of

63   *A terrestrial and celestial globe set by Richard Cushee (fl 1760).*

astronomy, such celestial globes were hopelessly inaccurate they were widely welcomed. The fact that distances of stars were not measured until 1838, and it was assumed that the brightest stars were the nearest, did not affect the validity of the celestial globes.

Of more use in the teaching of simple astronomy than celestial globes were armillary spheres and orreries. Armillary spheres are models consisting of rings or bands of metal encircl-

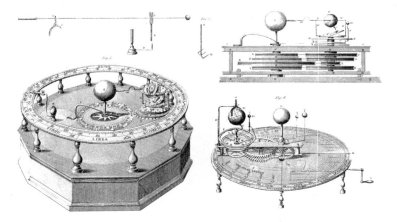

64   Diagram of an orrery, showing the mechanical parts.

ing a sphere representing the earth, and the first were made by the ancient Greeks, though these were fitted with sights and used as astronomical instruments rather than for demonstration and calculation. The bands represented the paths of heavenly bodies; some moved, usually the sun and the moon and occasionally other planets as well, while most of them were fixed, and were either purely ornamental and engraved with the signs of the zodiac or represented planets or constellations.

Armillary spheres were extremely popular in the sixteenth and seventeenth centuries, and although similar in shape their ingredients varied, depending on the ability of the maker to manufacture moving parts. Armillary spheres were mainly of brass, though the central sphere representing the earth could be of brass, wood, or glass, and varied in size, the larger wooden spheres being true miniature globes with the continents and the oceans delineated.

Celestial globes antedated terrestrial globes, which had to wait until it was accepted that the earth was round. Some globes had gadgetry fitted, mainly in the form of small spheres on the end of wires representing the sun and the moon, which were operated by clockwork and in some cases actually informed

65 *An orrery by B. Martin c1770, sold recently for £2,800.*

observers of the phases of the moon. Other globes had a good deal of somewhat unnecessary brasswork, with bases suppled with screws to level the globes as in the manner of theodolites, or with rack and pinion to tilt the globe.

The craze for armillary spheres died out about 1700, though globes with lunar and solar attachments were made until a much later date. The successor to the armillary sphere was the orrery, the first of which was made by the clockmaker George Graham (1673–1751) and his famous uncle, the doyen of clockmakers, Thomas Tompion about 1709. This was copied by John Rowley about 1712 for Charles Boyle, fourth Earl of Orrery, Rowley's patron and thus a suitable person to have an instrument named after him.

The original orrery was a furnishing piece rather than a scientific instrument, with intricate brasswork and a wealth of lacquered designs, and it depicted the relative motions of sun, moon and earth. The machinery was concealed on the drum-like base, and operated by a handle. Around the rim of the instrument ran a brass ring with a graduated scale, and a pointer

SPHERES AND ORRERIES

66 *A globe worked by clockwork, with the earth surrounded by a transparent celestial globe.*

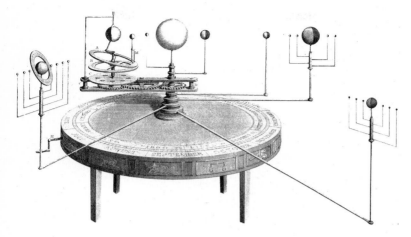

67 *A traditional orrery (or planetarium), with the machinery concealed in the drum-like base.*

moving with the revolving top indicated the date. The original orrery was pure clockmaking, calling for no great innovations, but succeeding orreries were considerably more complex.

The premises of John Rowley at the sign of the 'Globe' in Fleet Street were taken over by Thomas Wright. The demand for orreries as astronomical toys and conversation pieces caused him to change the name of the shop to the 'Orrery and Globe', and he acquired royal patronage. Wright incorporated the movements of the planets into his orreries, and although the mechanics were relatively simple compared with the marvellous parts of automata then being made on the continent they commanded wonder.

It is indicative of the close-knit world of instrument makers that Wright was succeeded by the maker Benjamin Cole, who in turn was followed at the same premises by the Troughton brothers. About the 1760s the making of orreries entered a new phase, and in place of flat metal plates carrying the spheres around their various symbolic routes Benjamin Martin invented an orrery in which the spheres were connected by brass arms

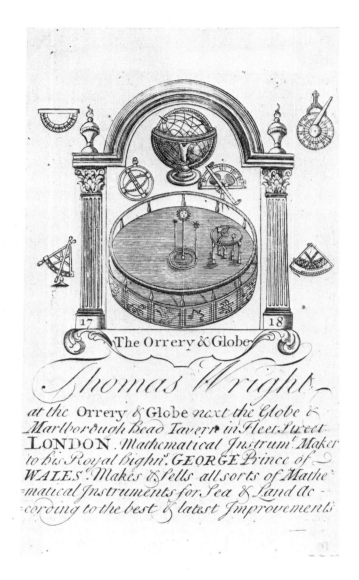

68 The trade card of Thomas Wright at the sign of the 'Orrery and Globe'.

to a central vertical rod, on which were fitted a series of graduated tubes, revolving at different speeds. Each arm was fitted to a specific tube. Other orreries by Benjamin Martin incorporated quite a complex system of gearing. The planets were still operated by rods to tubes on a central rod, but extra fitments were provided for the earth, so that its revolution was accomplished by machinery not involved in the movements of the planets. Extra gearing was run off the earth machinery to illustrate the movement of the moon.

Benjamin Martin called his orreries planetariums, and one sold recently in a London auction room for £2,800.

Several excellent orreries were made in the United States in the late eighteenth century. The premier American instrument maker, David Rittenhouse (1732–96) built one for the University of Pennsylvania, which was unusual in that it was constructed vertically, and set in a fine cabinet. Rittenhouse's orrery consisted of three main components, a large centre-piece illustrating the movements of the planets and two smaller pieces showing the eclipses of the sun and the moon. A more spectacular orrery was made by Joseph Pope (1750–1826) of Boston in 1787 for the University of Harvard, which did not have the funds to pay for it, and the orrery was sold by public lottery. Pope's orrery, which was $6\frac{1}{2}$ft high and also had a diameter of $6\frac{1}{2}$ft, was a masterpiece of its kind, though somewhat less complex and sophisticated than Martin's planetarium.

The orreries of the late eighteenth century were not only ornamental prestige pieces, but were well enough made to help in astronomical problems without the trouble of complicated calculations. Naturally the distance between planets was not represented to scale. Variations of orreries were also made to instruct in more specific matters, and Benjamin Cole, who succeeded Wright at the sign of the 'Orrery and Globe', made, for the benefit of the Fellows of the Royal Society, a model illustrating why the transits of Venus occur only at certain intervals (twice every century, with an interval between each transit of eight years).

The sterling efforts of the astronomers made it possible towards the close of the eighteenth century for maps to be made of the moon, and it was only a short step to constructing a moon globe. One of the most interesting was made by John Russell, who incorporated a small rotatable sphere representing the earth attached to the stand. He called this combination a selenographia. There were also terrestrial globes made to illustrate the precession of the equinoxes, globes based on the nineteenth-century 'maps' of Mars (which were wholly fictitious), while the fashion for illustrating popular astronomy by models continued in 1910 when Halley's comet made its predicted appearance.

Most globes seen today in sale rooms and antique shops are, of course, Victorian, and because they are excellent furnishing pieces they obtain fairly high prices. This is beginning to apply as well to globes made for the schoolroom. It would perhaps be overstating it to say that orreries are priceless, but it must not be overlooked that such objects are extremely vulnerable to the effect of old age, especially the spheres and the rods, and it is possible that there exist the mechanical movements of orreries, unrecognised and almost unrecognisable, in some damp attic or workshop, waiting for some knowledgeable collector to come along and claim them for his own. There are in fact, two dismantled orreries in Plymouth, much coveted by dealers in scientific instruments.

There is little danger that pocket globes will not be recognised, for these are extremely charming. A terrestrial globe is contained in a spherical case, usually covered with shagreen, while on the inside cover of the case is pasted a map of the sky. The miniature terrestrial globe derives from a 12in globe by John Senex made in the early years of the eighteenth century, but other makers of large globes also made pocket globes, including Senex himself, Cushee (fl1708–32), Nathaniel Hill, and, early in the nineteenth century, the globe maker Newton. Some idea of the interest in pocket globes may be gained from the price range of these globes — from £60–£150.

CHAPTER 6

SPECTROSCOPES

THE FIRST recorded use of the word spectrum was in 1671. The spectrum is the coloured band into which a beam of light is decomposed by means of a prism or diffraction grating. First discovered by Newton, the phenomenon that white light split up into colours intrigued scientists. When it became known that light emitted by different substances decomposed into different spectra there was great excitement, because it was possible to investigate the chemical nature of substances independent of their distance.

These possibilities appealed most of all to astronomers and those interested in the heavens. The science of the motion of the stars was being very expertly explored by Galileo and his successors, and now added to this was the science of the nature of the stars. Unfortunately it was not so uncomplicated as the first masters of spectrum analysis thought. In the solar spectrum, for instance, there appeared inexplicable dark lines, and the beautiful simplicity of the first experiments that indicated that every element had its characteristic and invariable spectrum was soon seen to be erroneous. It eventually became apparent that elements as a rule possessed more than one spectrum, depending on the physical conditions under which they became luminous.

Spectrum analysis, renamed spectroscopy by Arthur Schuster in 1882, fascinated many of the most prominent scientists of the early nineteenth century, including David Brewster and John Herschel, and although investigators travelled along many false paths much was discovered. In 1855 an American, David Alter of Pittsburg, published a paper in which he described the spark spectra of hydrogen, air, and other gases. By an ingenious series of experiments it was found that the dark lines of the solar spectrum arose from the presence of sodium in the gaseous atmosphere of the sun, and what was true of the sun applied also to the stars. A new era in astronomy began, and many authorities compare the implications of spectrum analysis in astronomy with the invention of the telescope.

A key figure in spectrum analysis was Joseph Fraunhofer (1787–1826), one of the greatest Germans in optics, who, by questioning the theory behind Dollond's achromatic lenses and improving on them, produced some of the finest object glasses of the early nineteenth century. He was responsible for the aplanatic lens (a word coined in 1827), a system which, for a given object distance, was free from spherical aberration.

Fraunhofer was the first man to examine and map the absorption lines (the mysterious dark lines) of the solar spectrum. He placed a flint glass prism before the object glass of a small theodolite and fed both with sunlight from a vertical slit in the window shutters of his room, and saw that the sun's spectrum was crossed by a series of vertical dark lines, which always kept the same relative order. About 1815 he constructed a $4\frac{1}{2}$in refracting telescope fitted with a prism. He accentuated the pattern of the dark lines by placing before the telescope objective a series of equally placed thin wires. Using his findings in terms of the wave theory of light, Fraunhofer was able to determine the wavelength of light with a good deal of accuracy. He and his pupil made a drawing 8ft long of the solar spectrum, and later scientists, such as Brewster, worked on this. This map led eventually to the definitive map published in Upsala in 1868 which gave the position of some 1200 lines, of which 800 were identified with the lines of common elements.

Astrophysics provided a new challenge for opticians, and John Browning, William Simms and the Hilger brothers of London, with Howard Grubb of Dublin, became acknowledged

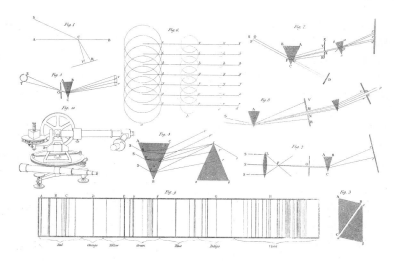

69   *Diagram illustrating the dispersion of light and a spectroscope.*

70   *A simple spectroscope with one prism by John Browning.*

leaders in the design and production of prisms and spectroscopes.

Despite the complexity of the subject, spectroscopes themselves are not devastatingly complicated, consisting of three main parts, the collimator (a tube with a slit and lens which collects the light and throws it upon the prism), the prism and its accessories, and a telescope. The slit of the collimator is adjustable. It confines the light, while the lens makes the beam parallel. This parallel beam is passed into the prism or chain of prisms. Trains of as many as fifteen prisms were used for those spectroscopes dealing with the sun, but for spectrum analysis of the stars fewer were used (because of lack of light). In modern astrophysics, the telescope is replaced by a camera. The telescope is fitted with cross-wire and other devices for measuring spectrum-line positions.

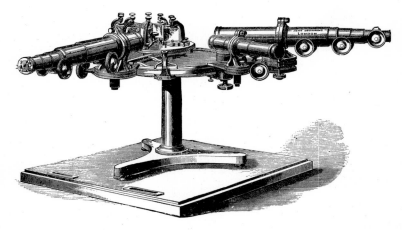

71 *A more sophisticated spectroscope with a train of prisms.*

One of the most important astrophysicians of the nineteenth century was William Huggins, who did valuable exploratory work from 1860 to 1869. Huggins used a cylindrical lens to give width to a star's spectrum, and as early as 1863 Huggins incorporated a camera into his spectroscope. In 1864 Huggins made an important breakthrough in astrophysics; he directed his attentions towards a planetary nebula in Draco, but to his disappointment there was no spectrum, only a single bright line. At first he thought that there was something amiss with his instrument, but soon the reason dawned on him. To quote his words: 'The riddle of the nebulae was solved. The answer, which had come to us in the light itself, read: Not an aggregation of stars, but a luminous gas.' In 1868 valuable work was done on the make-up of solar prominences with the aid of the spectroscope, and in 1870, also by the use of the spectroscope, the element helium was 'discovered', not to be isolated from a terrestial source for another quarter of a century.

The replacement of the telescope of a spectroscope by a camera initiated the spectrograph, first used in 1875, for explor-

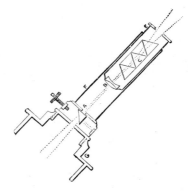

72 *Diagram illustrating the microspectroscope.*

ations in the ultra-violet range. The camera part of the spectrograph is in the form of a tube, elongated so that the bottom, holding the photographic plate, is rather wider than the top, nearest the prism. With this spectrograph Huggins succeeded in obtaining spectrograms of Venus and Jupiter, a comet tail, and the Orion nebula. The lenses and the prism were provided by Adam Hilger, who also supplied the optics for a more sophisticated spectrograph in 1880, incorporating a small red lamp *inside* the tube.

Much important work in spectroscopy was carried out in

America, and between 1886 and 1889 W. H. Pickering, director of Harvard Observatory, made a complete spectrographic survey of the stars in the northern hemisphere, resulting in a library of several hundred thousand spectrograms, but Potsdam Observatory vied with America in this productive field. Potsdam claimed to be the finest astrophysical observatory in the world.

The spectroscopes of the great observatories of the world demonstrate one side of the picture. On a more mundane level spectroscopes were used in laboratories in conjunction with a Bunsen burner. It had long been known that certain substances imparted different colours to flames. In 1822 Sir John Herschel described the spectra of strontium and copper and other substances, stating that 'the colours thus communicated by the different bases of flame afforded in many cases a ready and neat way of detecting extremely minute quantities of them'.

A few years later the pioneer of photography Fox Talbot described a method of obtaining a monochromatic flame by using diluted alcohol in which a little salt had been dissolved in a spirit lamp. In a paper he declared that 'a glance at the prismatic spectrum of flame may show it to contain substances

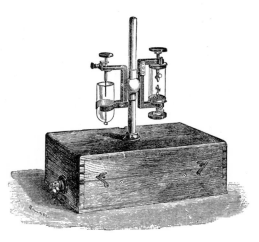

73 *Apparatus for spark spectra.*

which it would otherwise require a laborious chemical analysis to detect'. The spectroscope could be an amateur's toy as well as an astronomical instrument. Salt gave a bright yellow, potassium gave red plus violet, while barium salt produced some vivid greens. The spectroscope became the sophisticate's Kaleidoscope, an amusement — the ash of a cigar, when moistened with hydrochloric acid and held in a platinum wire in the flame of a Bunsen burner, informed the curious, by way of the spectroscope, that it contained lithium, sodium, potassium and calcium.

To cater for the general public, the instrument maker John Browning, prominent in fine spectroscopes for astronomers, produced a small direct-vision spectroscope retailing at 22s. This resembles a stubby telescope in form.

There are certain substances that call for a greater degree of heat than that produced by the flame of a Bunsen burner, and electricity was used, provided by batteries. Advantage was taken of the fact discovered by Faraday that an electric spark was nothing but highly heated matter, and Browning invented an apparatus for spark spectra. With the use of high temperatures the spectra of such substances as iron were discovered.

Browning also made a micro-spectroscope for the amateur. The eyepiece of the microscope was replaced by the micro-spectroscope, comprising an adjustable slit, a reflecting prism, and a train of five prisms for dispersing the rays. No doubt such instruments provided enjoyment for the dilettante, dimly aware that he was involved in mysterious scientific progress.

CHAPTER 7

MICROSCOPES

THERE ARE a number of interesting parallels between the telescope and the microscope. Just as there are two basic types of telescope, the refractor and the reflector, so are there two types of microscope, the simple and the compound. Both genera had complicated optical problems due to the difficulty of producing lenses without chromatic or spherical aberration, and mechanical problems were solved much more easily than optical ones.

Many of the men involved in the making of telescopes also made microscopes, and the coming of the age of brass, replacing card, shagreen, leather, parchment and wood, marked the emergence of the modern instrument. Until brass was predominantly the material used, telescopes and microscopes suffered in such matters as the question of the sliding tube.

The simple microscope consists of a single positive lens, or of a lens combination acting as a single lens, positioned between the eye and the object so that it presents an enlarged image. The compound microscope usually consists of two positive lens systems. The greatest advantage of the compound microscope is that it represents a larger area.

Like the Galilean telescope, the simple microscope was arrived at by accident. It is more akin to a magnifying glass as used by stamp collectors and watchmakers than the familiar microscope

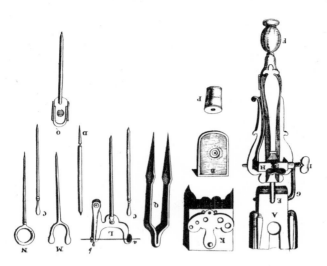

74 Diagram of a simple microscope, with accessories.

on a stand. The principle of the simple microscope rests on the phenomenon that any solid or liquid transparent medium having either one convex and one flat surface, or two convex surfaces whose axes are coincident, serves as a magnifier. The condition is that it shall refract the rays passing through it to cause widely diverging rays to become either parallel or only slightly divergent. Two convex surfaces are more powerful than one convex and one flat. If a tiny object is placed on a piece of glass and a drop of water is placed on it, it will be magnified. If a small hole is made in a strip of metal and a minute drop of water is placed in it, this spot of water, having two convex surfaces, will serve as a more potent magnifier. This method of making a simple microscope was used by late Victorian peddlers to make toys. Instead of water they used a drop of amber, and their microscopes were sold for a penny each in the London street markets.

The curious magnifying powers of a globe of water were known to Seneca in about AD 100, and in his archaeological excavations between 1845 and 1857 near the River Tigris,

Layard found a convex lens of rock crystal on the site of the palace of Nimrud. Certainly the superb gem cutting of the ancients could not have been accomplished without the use of magnifiers.

The first illustration showing a magnifying glass in action was made in 1592 by George Hoefnagel. In his *Dioptrique* (1637) the philosopher Descartes describes microscopes using a concave mirror in conjunction with a lens, following on from Roger Bacon (1210–94) and d'Armati, the so-called inventor of spectacles. However it was a Dutchman, Antony van Leeuwenhoek (1632–1723) who first systematically drew the attention of the world to what the simple microscope could accomplish, making a lens with magnification of 270 and confusing science with his discoveries of bacteria.

The simple microscope existed in a variety of forms. The Leeuwenhoek microscope consisted of a single bi-convex lens mounted between two brass plates rivetted together. The microscope of Johan van Musschenbroek was more sophisticated. The lens was linked up with a device for holding the object to be examined by means of ball and socket joints, making focussing easier and holding the object firmly so that it could be systematically examined and drawn. A variation of this was the compass microscope, introduced towards the end of the seventeenth century, consisting of a pair of compasses, one leg carrying the lens, the other the object, placed or impaled in forceps or on a point. A fine screw between the legs made focussing easy; the compass microscope was held in the hand by a handle.

Focussing was expedited even further by the screw-barrel microscope associated with the name of James Wilson (1665–1730), made in two parts, a barrel with the lens at one end, and a barrel fitted with two sprung metal plates between which a specimen holder was pushed. The screw-barrel microscope, more like what is commonly thought of as a microscope than the others, had a condensing lens at the far end. The screw-barrel microscope could be held up to the light to get greater illumination.

75 *A barrel microscope by Campani (?1686).*

Sufficient illumination was a problem for the early microscope users, and an ingenious method to solve this was what was known as the 'Lieberkühn' devised by Descartes in 1637, a concave reflector with an aperture in the middle of the lens adding extra light on an opaque object.

Although the first compound microscopes were constructed between 1612 and 1618, it was for a long time believed that compound microscopes would never be as good as simple microscopes, and therefore research on simple microscopes, improving the optics by using lens combinations, was carried out for much longer than one would have thought. These combinations included doublet lenses. Perhaps the most important of these systems was that devised by William Wollaston (1766–1828), who not only distinguished two new metals, palladium (1804) and rhodium (1805) but succeeded in making platinum malleable, a process that earned him £30,000.

Wollaston used a combination of two plano-convex lenses of different focal lengths, plane sides towards the object and the smaller of the two lenses nearer the object. This lens system was further improved by inserting a diaphragm between the

76 *A simple pocket microscope in a shagreen case.*

two lenses. Other lens systems for the simple microscope were those by Fraunhofer (1787–1826), two opposing plano-convex lenses thus presenting a plane surface to both the object and the eye, and Steinheil and Chevalier (1794–1873) who used triple lenses.

Although it is certain that compound microscopes were produced early in the seventeenth century, no trace remains of these instruments. The development of the compound microscope depended on finding a suitable object glass. To a certain extent astronomical telescopes could get by with chromatic and spherical aberrations, but the problems were greater in the microscope. A compound microscope by Cornelius Drebbel was allegedly demonstrated in London in 1621, eleven years before Leeuwenhoek was born, but nothing really significant happened until 1665 when Robert Hooke published his *Micrographia*.

In a preface he described the instrument that he had been using. The optics comprised the objective, the field lens, and the eye lens. The barrel of the microscope was composed of four draw-tubes, and was supported on an upright pillar along which a wooden ring could be slid and held at the optimum height by a clamping screw. This ring was connected with a second wooden ring by means of a ball and socket joint, into

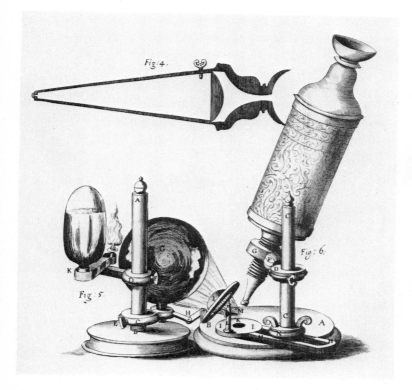

77 *The Hooke microscope, in which all the elements of the modern microscope were gathered together for the first time.*

which the barrel of the microscope was fitted. The tubes were made of cardboard. The basic elements of the modern microscope were here, except that the specimen, affixed on a spike or presented on a disc, was illuminated by a system consisting of a lens, a globe of water, and a lamp, which was not part of the microscope. Later this cumbersome appendix was replaced by a mirror beneath the stage (where the specimen is positioned).

Hooke was not the only one in the field, but it was he that made microscopy respectable. That compound microscopes were not merely one-offs is clear from a reference to the instrument maker Reeves (also prominent in telescope making) who was selling compound microscopes for between £3 and £6 in 1662. Unfortunately many microscope makers of this period did not sign their instruments, though two existing compound microscopes in museums have been assigned to the Italian maker Divini. The Italians were very prominent in the seventeenth-century microscope trade. Guiseppe Campani (1635-95) is considered to have been the first to use the screw-barrel as a focussing device, Divini (1620-95) the first to incorporate a doublet lens in the eyepiece.

Wood and cardboard were the basic materials of these compound microscopes, though brass was being used for bases, and of course the vulnerability of cardboard and wood is unquestionably responsible for the fact that few instruments of this period exist.

Despite improvements in the mechanics, the optics were still faulty. Microscopes became beautiful objects, amateurs' instruments rather than tools of discovery and knowledge. English makers, such as James Mann (1660-1730) took their design lead from Italy, using a tripod stand and the screw barrel principle. John Marshall (1663-1725) broke away in 1693 from the Italian emphasis on the vertical. The pillar was mounted on a ball and socket joint so that the barrel of the microscope could be tilted. The stage holding the specimen was attached to the pillar, so that tilting the microscope body would not throw the specimen out of focus. No longer did the base (in this case of solid wood) have to be beneath the specimen. By tilting the body, a candle or a lamp could be placed *beneath* the specimen.

Marshall was one of the first to use a fine adjusting screw—a long metal screw moving the body up and down as the nut is turned. Sliding tubes had been introduced in Italy about

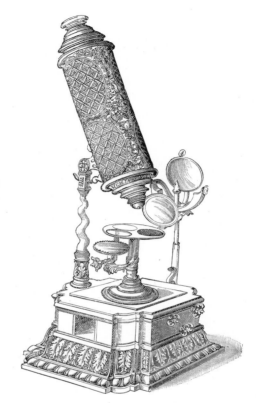

78 A microscope of c1716.

1667, rack and pinion had been introduced in Italy in 1691 for focussing, and with the fine adjusting screw by Marshall, who took his lead from astronomical instrument makers, all the mechanical elements of today's microscopes were present.

Nevertheless, the Marshall instrument was only one of many types on the market at the turn of the century. Details of a microscope with six lenses were published in 1667, and binocular microscopes were being devised, while to get over the inadequacy of the objectives there was a good deal of investigation into spherical lenses. Spherical lenses were a short-lived fad, and although some were fitted into simple microscopes they did not live up to expectations.

MICROSCOPES

79 John Marshall was one of the first to use a fine adjusting mechanism, and provided a wide range of accessories.

Excellent as were Marshall's innovations, the ball and socket joint and his fine focussing screw put the prices of his instruments up. The ball and socket principle was ignored by other makers in preference to the method made popular by Edmund Culpeper (1660–1730), an instrument maker of Moorfields. In 1725 Culpeper introduced his vertical microscope set on a double tripod. On top of the first tripod is the stage, perforated to allow light from the concave mirror situated on the second tripod. This mirror could be swivelled, so that sufficient light was at last available to the user without the unwieldy illumination system evolved by Hooke. A condenser lens mounted on the upper tripod provided light for opaque objects.

80 Excellent as Marshall's innovations were, they pushed the prices of microscopes up, and many preferred the double tripod instrument of Culpeper.

Culpeper made a good many microscopes of this type, far more than Marshall, and the general design was popular for a century, though the wood and cardboard were replaced by components made of brass. Microscopes of the early eighteenth century were equipped with a range of accessories, the most important being extra objectives of varying powers (usually ×30 to ×200). Yet even at this stage simple microscopes were in most respects more efficient optical instruments than compound ones, and Culpeper and his contemporaries were still turning them out, usually of the screw-barrel type.

The great advantage of the Culpeper double-tripod instrument was that it was sturdy, simple and stable. In utility however, it was eclipsed by the Cuff microscope. John Cuff (1708–72) looked at both the Culpeper and Marshall instruments, and also saw that microscopes such as the Martin drum microscope (the entire instrument in one barrel with cut-outs to give light) were obsolete. His microscope incorporated a brass vertical pillar, a better fine-focussing device than Marshall's, possible when all was made of brass except the wooden base containing a drawerfull of accessories, an open stage (the main defect of the Culpeper tripod system was that access to the stage was cluttered) and a very large gimbal-mounted mirror beneath the stage. Coarse focussing was done by the rack-and-pinion method, and unquestionably the Cuff microscope was the most efficient instrument to appear on the market; it was unashamedly copied throughout Europe. Cuff himself made

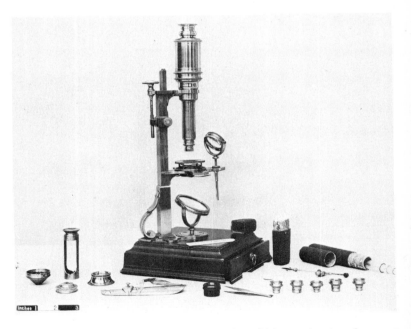

81 *The microscope of John Cuff (1708–72) was by far the most efficient microscope of its day.*

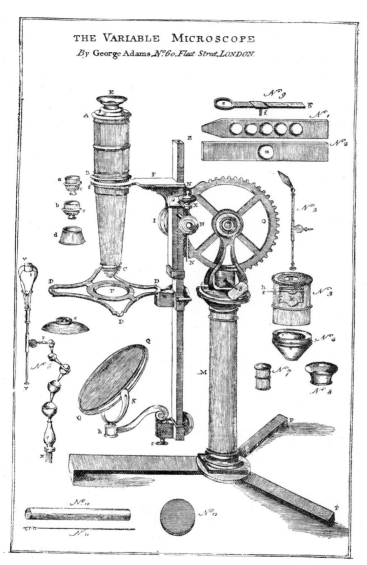

82 *The microscope of George Adams.*

adaptations, adding in about 1747 a micrometer made up of a lattice of silver wires, and an interesting modification was the moving of the stage and not the tube for focussing, which occurred about the same time. The instrument maker James Ayscough brought in in 1749 a sliding objective-holder. This obviated the need to unscrew the object glass when a change of power was needed. The objective-holder was a rectangular brass plate with four holes into which the objectives were screwed before use.

Not surprisingly the inclining pillar of the Marshall microscope was seen to be an admirable feature, and with brass being now the predominant material used the ball and socket joint was replaced by a hinge, and among the makers of the mid-eighteenth century who drew from Marshall and Cuff their best innovations were Watkins and Nairne, Martin and Shuttleworth, though it was the German Tiedemann who consigned the wooden base to oblivion in favour of a flat brass tripod, a method of establishing stability that remained in favour until well into the nineteenth century.

In 1758 John Dollond put on the market his achromatic combination for telescope objectives, but making such lenses for microscopes was a different proposition to making them for telescopes, and it was not until 1830 when V. and C. Chevalier produced objectives that makers really capitalised on the achromatic breakthrough.

The making of achromatic objectives for microscopes was complex and expensive. About 1824 the Frenchman Selligue had endeavoured to make high-power objectives by screwing together several low power (therefore more easily ground) achromatic lens combinations, but these produced cumulative aberration that was not understood until in 1830 Joseph Lister (1786–1869), the father of the surgeon, laid down the theory, and by using this data Andrew Ross simplified the whole procedure in 1837 by using what was known as a correction collar. This method was further improved by Wenham and Zeiss. Andrew Ross was a key figure in early Victorian optics, and in about 1860 he and Wenham introduced the binocular microscope to the public.

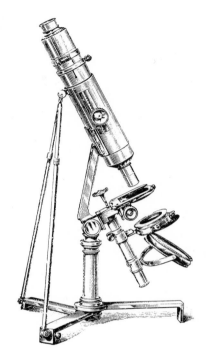

83 Tully's achromatic microscope c1826.

After the death of John Cuff in 1772 there seemed little that instrument makers could do to improve the mechanics of the microscope. The microscopes made for George III by George Adams are among the most beautiful instruments of any kind ever made. There were interesting varieties of microscope, such as the solar, a relatively unsophisticated hand-held instrument in which the light was provided by the sun via a large mirror. Museum microscopes, probably invented by Thomas Winter of Brewer Street, Golden Square, were also popular. The main feature of the museum microscope was the revolving ivory stage, in which there were cut-outs for specimens. As many as 65 pre-selected specimens could be arranged in the apertures, and

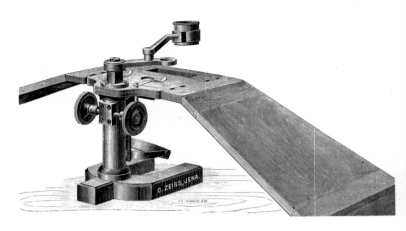

84   A Zeiss dissecting microscope.

it was ideal for teaching purposes, though the instruments lacked the fine adjustments of the Cuff and Marshall microscopes. The cut-outs were placed in rows, and the instrument swivelled on the stand so as to encompass all the specimens. An extremely sophisticated development of the museum microscope was carried out late in the nineteenth century, called the multi-ocular microscope, five microscope tubes ranged in an arc about a 90° prism. Although there were five eyepieces there was only one objective, and the prism was used to direct light from this objective into any one of the tubes.

An intriguing attempt to make a microscope unvexed by optical problems was made by G. B. Amici (1785–1863), and his reflecting mirror slightly antedates the first effective achromatic lens microscopes. The reflecting microscope is the exact analogy with the reflecting telescope, and both were brought into being through the same cause. By using an elliptical mirror he considered that he would circumvent all the optical problems roused by lenses, though Amici is more important for his role in investigating what were known as immersion lenses. Water immersion was used in 1840 by Amici, but in 1878 Ernst Abbé and Carl Zeiss improved on water, using Canadian cedarwood oil.

Although the ultimate in lenses did not come until 1886—the so-called apochromatic objective of Ernst Abbé—this does not mean that microscope design stood still. In 1841, Cornelius Varley (1781–1873) demonstrated his lever stage movement, enabling the stage to be moved in a horizontal plane in any direction, so that living creatures could be tracked. He used the instrument makers Powell and Lealand to make these microscopes for him, and these are among the finest instruments made during the Victorian period. This firm lasted from 1841 until 1914.

There are two other developments in nineteenth-century microscopy that deserve mention. The less important was the inverted microscope, made for viewing objects from below. A slanted mirror carried the image from the vertical tube housing the objective to the tube containing the eyepiece. This tube was fixed at 45° angle to the vertical tube. It was an ingenious microscope made for a limited purpose, and does not compare in significance with the binocular microscope.

There was a little seventeenth-century speculation in binocular vision, but there matters rested until Charles Wheatstone, a pioneer in the field of the electric telegraph, interested himself in stereoscopes. After the Great Exhibition of 1851 stereoscopes became a popular entertainment.

Although the American Riddel is credited with the first binocular microscope in 1851, Ross and Wenham were responsible for its acceptance. A prism was mounted covering half of the objective. This deflected the rays into the second tube, fixed at an angle to the vertical tube. The uncovered half of the objective formed the usual image in the vertical tube. When viewed through the eyepieces the two images merged into one. Binocular microscopes were very useful for observing three-dimensional objects, and are unquestionably less tiring for

85　*An aquarium microscope, really an augmented magnifying glass.*

86　*There are many fine Victorian binocular microscopes available at reasonable prices. These date from 1860–80. The one on the left made £68 at auction, the other £140.*

researchers who constantly have to resort to their microscopes.

The standard of Victorian instrument making is unquestionably very high indeed, and this is especially true of their microscopes, in which the ornament and fussy decoration so often noted not only in furniture and applied art but in such things as telegraph equipment and typewriters is totally lacking. In form and function the Victorian microscope is supreme, and it is a pity the attitudes of their makers towards their products did not infiltrate more into everyday life.

CHAPTER 8

BAROMETERS

A BAROMETER is an instrument by which the weight or pressure of the atmosphere is measured, but came into being from attempts to find out if a vacuum was possible, and only secondarily 'to make an instrument which might show the changes of the air, now heavier and coarser, now lighter and more subtle. Many have said that the vacuum cannot happen . . .'. So wrote Evangelista Torricelli (1608–47), the pupil of Galileo and a mathematician and a maker of optical lenses. Galileo observed that a common suction pump could not raise water to a greater height than about 32ft, and could not see why. Torricelli could, if atmosphere had weight and the pressure which it exerted was equal to that of a 32ft column of water.

He then reasoned that if the atmosphere supported 32ft of water it should support $2\frac{1}{2}$ft of mercury, mercury being about $13\frac{1}{2}$ times heavier than water. He tried this out by having a 4ft glass tube made with one end closed, which he filled with mercury. Closing the open end with a finger tip, he dipped the tube, keeping his finger over the opening, in a bath of mercury. He released his finger. The 4ft of mercury in the tube instantly sank to 30in, leaving in the top of the tube an apparent vacuum. His predictions were fulfilled, and this trial became known as the Torricellian experiment. It received great attention.

BAROMETERS

The philosophers Descartes and Pascal considered that this device would be useful to ascertain the heights of mountains. At the top of mountains the atmosphere would diminish, and the column of mercury would fall. Pascal went up and down the highest Paris towers, his brother-in-law toted the new-born barometer up a mountain, and it seemed that a valuable aid to surveying had come into being. Only when it was discovered that the rise and fall of the mercury was also affected by weather changes was there any rethinking, and thus the instrument began to be called a weather-glass.

The barometer is a simple object, and did not need specialised workmen to make it. There are two basic types of liquid thermometers (mercury was often used but not always): the cistern barometer and the siphon barometer. The cistern barometer is no whit different to the original Torricellian apparatus — a tube sealed at one end and with its open end in a container of mercury. The siphon barometer dispenses with the cistern, and employs a U-tube with one leg larger than the other. The reading is taken of the difference in the levels of the mercury in the two legs.

Before the end of the seventeenth century barometers did not always give consistent readings, and few makers dared to use a graduated scale, preferring to indicate *changes* in the weather. There was no problem in getting the tubes, but there was in filling them with mercury without leaving any trace of air or vapour. Not for many years did it occur to anyone to boil the mercury, and thus eliminate air and vapour. The discrepancies in mercury levels were thought to be due to porous glass, and that air was getting in and upsetting the readings.

It was an enigma to philosophers. If air could pass through the glass in small quantities, what was there to stop air getting into that part of the tube in large quantities and sending the mercury right down the tube into the cistern? Therefore it was not air from outside that was in the tube. When the tube was inclined the mercury sped into the empty part of the tube; therefore the empty part of the tube did not contain air. So philosophers postulated an 'ether' or a 'subtle matter'; to them, the barometer was a curious laboratory experimental apparatus.

87 The barometer was associated with laboratory experiments attempting to create a vacuum. Pneumatic machines using an air pump were rare one-offs, but they may occasionally be encountered.

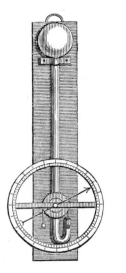

88 A barometer with a calibrated wheel, invented by Robert Hooke.

As a commercial apparatus the seventeenth-century barometer left much to be desired. The inherent fragility of glass tubing and a dish of mercury led to them being mounted on a board. The tiny variations in the height of a mercury column were difficult to perceive. Robert Hooke used a dial consisting of a pointer moving in front of a calibrated brass wheel, the pointer being set moving by a thread passing over a small pulley on the ends of which were small spheres. One of these spheres rested on top of the mercury in the shorter of the two U-tubes. Movement of the mercury level therefore raised or lowered the sphere, which by means of the pulley activated the pointer.

This method was described by Hooke in 1665; in 1668 he used two liquids (mercury and water) in a U-type barometer incorporating two bulbs. The mercury was the prime force, and as the water was in a narrow tube variations in the mercury level were therefore much magnified. The difficulty in making these double barometers was in getting glass tubes of the requisite thinness. Both the wheel and double barometers were extremely popular, but another method to make for an easy reading was a diagonal barometer (1668). The mercury level, on a slant, moved more perceptibly along a diagonal.

Yet another answer to the small variations in the mercury level in ordinary syphon and cistern barometers was furnished by Amontons, who produced a folded barometer in 1688 consisting of a number of parallel tubes connected alternately at the top and the bottom, the alternating tubes containing mercury and 'red liquid'. In 1695 he invented a conical barometer that operated from the top down—the mercury went down the tube instead of up it. None of these gained popularity, and even the Hooke double barometer was distrusted, and was often combined with an ordinary barometer as a check. The greatest disadvantage of the double barometer was that the tube containing the lighter liquid got dirty where the surface of the liquid rose, and to counter this Hooke invented a three-liquid barometer. This innovation was claimed by the French, unsuccessfully, and it was possibly in response to this rebuff that

89 Portable barometers were used to ascertain heights of mountains and hills, and the tube was filled with mercury on the site.

Nicholas Gauger dreamed up the four-liquid barometer in 1710.

During the first two or three decades of the eighteenth century the diagonal barometer set on square mahogany boards became popular, but despite all the variations and the ingenuity of Hooke, the primitive Torricelli barometer remained the favourite, especially among physicists, who were happy to take an empty tube and a container of mercury with them into the field, and fill the tube on the site.

In the 1720s inconsistency in mercury behaviour was cured by having the mercury boiled in the tube, and by 1750 most makers of barometers adopted the idea. Barometer and thermometer makers had formed themselves into cliques, called themselves physicists, and had dissociated themselves from the instrument makers. At the start of the eighteenth century instrument makers began to make their entry into barometer making, including Francis Hauksbee, who made an L-shaped barometer with the mercury cistern at the top of the upright and called it a baroscope, but as with other 'upside-down' barometers such as Amontons' of 1695 it did not catch on. Nor did it when the

L-shape barometer was re-invented by T. Horton in 1898.

Compared with the multitude of tasks that the makers of optical instruments or complicated angle-measuring instruments had to accomplish, the manufacture of barometers was easy. There was almost as much difficulty in setting them; Jesse Ramsden conceived the idea of zeroing the mercury level by making fine adjustments to the size of the cistern by means of a screw through a wooden box and pressing the leather (and therefore flexible) bottom of the cistern. This was improved upon by means of a screw and a double layer of chamois leather, and Ramsden and his workshop made further modifications to alter the capacity of the cistern.

Perhaps a simpler way was to adjust the tube (1778) or move the index, but the zeroing of a barometer presented many problems to the makers.

The words on barometer scales have altered surprisingly little. In 1688 George Sinclair graduated his barometer 'Long Fair; Fair; Changeable; Rain; Much Rain; Stormy; Tempests'. The commonsense Victorians had similar labels, as per a Negretti and Zambra barometer of 1864: 'Very Dry; Settled Fair; Fair; Changeable; Rain; Much Rain; Stormy.'

Until about 1828 portable barometers were a chimera. Not only were barometer tubes fragile, but, far more important, minute quantities of air could enter the instrument during transportation without being noticed and thus render readings inaccurate. Most serious users preferred to take the components and fill the tube on the spot. The air trap invented by the Paris instrument maker Bunten solved the problem; it was a small modification of the barometer tube carried out at trifling expense.

A more drastic portable barometer was one with the tube made of iron. The first reliable iron-tube barometer was devised by the French mathematician J. de Blondeau in the 1770s. It was read by means of an iron wire projecting from an ivory float on top of the mercury which was noted against a scale. An extremely complicated steel barometer was made by Nicolas

Conté towards the end of the eighteenth century, but this was a one-off, and in any event, with the arrival of efficient angle-measuring instruments and the sophistication of the art of cartography the barometer as a height-measuring instrument was less in demand.

Ships needed barometers that were sturdy, portable, and did not mind being jigged about. The first true marine barometer was made by Edward Nairne (1726–1806) in about 1773. His barometer was kept in position in the same way that the ship's compass was kept steady in its binnacle, by having a gimbal mount. Nairne kept the barometer vertical by a weight fastened at the bottom of the instrument. His most important innovation was a long narrow (one-twentieth of an inch diam.) tube, and this became a feature of marine barometers. Compared with drawing-room barometers, increasingly in demand during the nineteenth century, marine barometers are slender and functional, usually carrying a thermometer alongside the barometer scale, and are much more collectable than the household variety if only for their quaint scale wording 'Rise for Cold, Dry or Less Wind—Except Wet from Cooler Side; Fall for Warm, Wet, or More Wind—Except Wet from Cooler Side', legends demanded by Admiral Fitzroy (1805–65).

By the early eighteenth century the barometer was a familiar instrument in rich homes and learned circles in Britain, but America did not see a barometer until 1727 when a London merchant gave Harvard University a collection of instruments. Native born barometer makers were handicapped by the absence of glassworks, and the earliest extant American barometer seems to date from 1796, a fine instrument made by Thomas Dring of West Chester, Pennsylvania, though it is clear from contemporary advertisements and trade cards that Dring was not the only man engaged in the making of barometers. In 1787 Joseph Donegany advertised in the *New York Daily Advertiser* that he made 'thermometers, barometers, and sold hydrostatic bubbles and hygrometers for proving spirits' and even earlier Joseph Gatty made and sold in New York 'every simple and compound form of barometer and thermometer'.

That barometers were rare in America is evident from a plaintive letter from the President of Yale in 1744: 'We have a very good Borometer (sic) and Thermometer both together, but the Borometer Tube is broke, and I believe no Man in this Country can put in another.'

There were barometers in America, imported from London, including instruments by Scarlett and Champneys but it was not until 1832 that they became readily available when an English instrument maker, James Green, established a business in Baltimore, specialising in barometers, though he could still not cater for the demand. The United States Navy had to order barometers from the British maker Adie, though later Green provided their requirements.

Although Green made several innovations in his barometers, including an improved cistern, other American barometer makers contributed remarkably little especially when one considers the progress made across the technological board. Patents by W. R. Hopkins of Geneva, NY, T. R. Timby, and L. Woodruff were taken up but nothing further was heard of them.

In Britain, the eighteenth century saw considerable advances in barometic accuracy. The index pointer was replaced about 1775 by an index ring which enclosed the tube, and a Vernier scale was added about 1770 to a thermometer by Ramsden; the Vernier became common in not only laboratory barometers but domestic barometers as well. It need hardly be mentioned that the cases of these late eighteenth-century barometers are in the highest tradition of British cabinet and furniture making.

An important innovation arose about 1850, when the entire scale was enclosed in a glass tube. An interesting variation appears about 1880 when the index was made to move by a screw mechanism. This type of barometer was marketed by Negretti and Zambra, but does not appear to have been very popular. To casual users of the barometer the addition of a microscope does not seem necessary, though this was mooted as early as 1698, and Sisson provided a barometer with a microscope for reading the Vernier. There have also been ingenious mirror gadgets for avoiding parallax, again surely unnecessary.

Late eighteenth- and early nineteenth-century barometer makers were very much indebted to Nicolas Fortin, who improved the barometer cistern. This cistern could be adjusted to the finest of limits and provided the perfect method of zeroing

90 Nicolas Fortin improved the barometer cistern, using a screw pressed against a leather container so that the surface of the mercury could be zeroed.

barometers. 'Fortin barometers' were made by most of the best British makers such as Adie, Negretti and Zambra, and Casella.

The Fortin cistern consists of a glass cylinder, through which the level of the mercury is seen, and below it a kind of leather bag, against which a screw works. At the top of the interior of the cistern is a small piece of ivory, the point of which coincides with the zero of the scale. By turning the screw, the leather bottom of the cistern is raised or depressed until the mercury surface is brought to the tip of the ivory.

The supremacy of the mercury barometer appeared in the opening years of the nineteenth century to be unchallenged but as early as 1698 the philosopher Leibniz was speculating about a 'little closed bellows which would be compressed and dilate by itself, as the weight of the air increases or diminishes'. Four years later he returned to the subject. He was thinking about 'a portable barometer which could be put in the pocket, like a watch; but it is without mercury, whose office the bellows performs, which the weight of the air tries to compress against the resistance of the steel spring'.

Without knowing it, Leibniz was anticipating the aneroid barometer by a century and a half. Nicolas Conté ventured into the same field in 1797, but his aneroid barometer was no more spectacular than his steel barometer, and it was left to Lucien Vidie (1805–66) to effect the breakthrough. Vidie started by making thin metal structures that held a vacuum, and in 1843 he made the first satisfactory aneroid (without liquid) barometer, and patented it in 1844, though it was not Vidie but E. Bourdon who marketed it under the name of manometer in 1849. Bourdon manufactured 10,000, but Vidie sued him for patent infringement and in 1858 Vidie was awarded 25,000 francs (£1,000) damages.

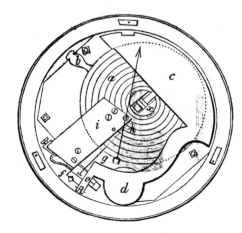

91 *A diagram of the aneroid barometer.*

Nevertheless, Vidie delivered more than 5,000 aneroids to Britain within a few years, and they proved very acceptable and were surprisingly accurate. An official of the Royal Observatory, Greenwich, tested one rigorously, and found that in a range of temperatures from 28° to 80°F error rarely exceeded a tenth of an inch. Patent rights in France expired in 1859 and the firm of Nauden, Hulot and Cie made 20,000 aneroids between 1861 and 1866. The leading makers of aneroids in Britain were Negretti and Zambra, who succeeded in producing a pocket-sized model in 1861.

The principle of aneroid barometers is simple; a flat circular box or chamber is partly exhausted of air, and acts as a spring which is affected by every variation in the atmosphere. The chamber is fixed to a level which operates, through a spring and other mechanisms, the pointer on the dial. There is usually a device introduced to counter changes of temperature, sometimes a bimetallic link in the mechanism, sometimes adjustment of the amount of air or other gas in the chamber. Occasionally aneroid barometers are extremely sophisticated, but most of

92 *A barograph, a recording aneroid barometer.*

these are modern, and the nineteenth-century aneroid barometers conform to a pattern. To many, it is rather a dull pattern, and collectors of aneroid barometers are few indeed, though it would be erroneous to suppose that all such barometers are of the cheap and nasty variety, an impression it is easy to get from the quantity of mass-produced aneroid barometers served up in cheap oak cases, the hallmark of lower-middle class Victorian respectability.

CHAPTER 9

THERMOMETERS

THE FIRST instrument to measure temperature was probably made in 1592 by Galileo. It consisted of a long thin glass tube with a bulb at one end; the bulb was heated to expel some air, and then the rod was up-ended and plunged into a vessel containing water or wine. When the bulb cooled, the liquid rose to a certain level in the tube, and subsequent variations in the temperature were shown by the rise and fall of this liquid caused by the contraction and expansion of air left in the tube. Inconvenient and solely a laboratory instrument, Galileo's thermoscope remained a curio until 1640 when the first liquid-in-glass thermometers were constructed.

The Italians were the first to persevere with the thermometer, and the Florentine investigators created an instrument consisting of a straight or spiral tube of glass, closed at the top and ending at the bottom in a bulb. The tubes were calibrated by globules of coloured glass attached to the outside of the tube. The graduations seem to have been arbitrary, varying enormously from instrument to instrument. The Florentines preferred alcohol to water as it reacted more sensitively. Contemporary writers declared that these thermometers were made 'rather for fancy and curiosity' than 'for any accurate deduction', and not until after 1700 were the temperatures of melting ice and boiling water adopted as fixed points for thermometer scales.

Impetus towards such fixed scales was provided by an influential book published in 1688 in Holland, in which thermometers with bulbs the size of a pigeon's egg surmounted by tubes 3 ft in length were described. Coloured alcohol was used in the tubes which were fixed to a grooved board on which graduations were marked. The fixed points used were the freezing point of water and the melting of butter. Budding thermometer makers were requested to divide the space between these two and mark *Tempéré*. From experience with barometers, it was believed that thermometers would only give comparable results if kept in the same place.

Little interest was shown in thermometers until 1717 when the Fahrenheit scale was introduced. The Réaumur scale came in 1730, and the Centigrade or Celsius scale in 1742, but it still took some time for a use to be seen in the thermometer. It would not assist surveyors, as the barometer did, and the sudden profusion of scales created confusion. There is no evidence that thermometers were produced on a commercial scale before 1750. The freezing and boiling points of water were key marks for the three major scales, but as late as 1775 there was dispute as to whether these were the best.

Although thermometers were produced by Robert Boyle, Newton, and Amontons, they were considered curios. In Britain, the Fahrenheit thermometer gained ascendance over its rivals; after considerable thought, Fahrenheit chose mercury as the filling for his thermometer tubes, Réaumur spirits of wine. Thermometers were not standardised; makers fixed the freezing and boiling points of water and filled in the gap between as they wished, and also those points on either side of the two major index marks.

A certain fillip to the thermometer trade in England was given by the interest of King George III in all scientific instruments. A collection of 'philosophical instruments' was given to him about 1760 containing a mercury thermometer made by Adams graduated from 10° to 129° Fahrenheit. George Adams's thermometer frames illustrated mid-eighteenth century confusion, for it is marked with four scales, Fahrenheit, Réaumur, Newton, and de Lisle. The de Lisle scale is interesting in that it runs in the opposite direction to most others, from 0 'Water boyles vehemently' to 160 'Just Freezing'.

The thermometer had from the first assumed a shape that we would instantly recognise, and variations from this norm were even in their day considered freaks. A French thermometer used a dial in the same way as Hooke used a dial in his wheel barometer; this thermometer incorporated an M-shaped tube. Drawing-room thermometers for the fashionable had various shapes of reservoir.

With the comparative lack of interest in thermometers before the middle of the eighteenth century it is surprising that clinical thermometers were in use by the 1740s, and although Fahrenheit was making clinical thermometers in Amsterdam, the doyen of the trade was the London maker, Alex Wilson. The early clinical thermometers consisted of a mercury-filled stem hermetically sealed inside a cylindrical glass tube 5 in. or so long, with the bulb at the base of the outer tube. The scale was written on paper, and inserted between the stem and the outer tube. Not until 1866–7 were thermometers systematically used in British hospitals.

The makers of thermometers encountered many of the problems met by the makers of barometers, and as many of the processes involved in making these two instruments were the same, makers of barometers were usually makers of thermometers, too. One of the chief difficulties was in acquiring standardised narrow-bore glass tubing, and where there was no good glass industry (as in the United States) research languished. It was no accident that thermometer making on any kind of serious scale began in Florence where there was glass-making experience.

By the beginning of the nineteenth century most of the problems associated with thermometer making were solved, and thermometers were used for a great range of activities. The advantages and disadvantages of the various liquids were realised. Mercury remained the most popular for use between −30°C and 500°C, and a twentieth century combination of mercury with thallium extended the range down to −55°C. Alcohol thermometers could measure down to −80°C, toluol thermometers down to −100°C and pentane yet further, down to −200°C.

Standard thermometers were produced of great accuracy which served as regulators for production models, and perhaps the most interesting of these was in use at Kew Observatory from 1851 onwards, $22\frac{5}{8}$in long and graduated in tenths of a degree from 0° to 75°C. This was replaced by a thermometer, made by Pastorelli and Rapkin of 46 Hatton Garden, which was 24in. long and had a slightly wider range. These ranges did not compare with portable standard thermometers, which were carried about in sets, each one covering a different range, but the most efficient regulator thermometers were made for the International Committee of Weights and Measures by the Paris makers, Tonnelet, in 1890, a set of three that measured from −39°C to 206°C.

Nineteenth-century thermometers included a wide variety of clinical thermometers, including ear and skin thermometers with an expanded bulb to present a greater area of skin to the instrument, earth thermometers with a long stem for the use of gardeners, thermometer sticks up to 3ft long for measuring ground temperatures, travelling thermometers, extra-sensitive thermometers with helical bulbs, jam thermometers encased in brass, and fluorescent thermometers using alcohol for use at night. The fluorescent thermometers were much used by the Arctic and Antarctic explorers such as Scott. Among the makers responsible for the unusual types of thermometer were Pastorelli & Co of 208 Piccadilly, Dring & Fage, Cetti & Co of 11 Brooke Street, and Griffin & Tatlock.

Industry was demanding high-range thermometers. The boiling point of mercury is about 357°C but by filling the upper part of the stem with gas under pressure, and using hard glass or quartz for the stem, this could be extended to 550°C, though mercury-in-steel thermometers were better still. Many of this latter type were designed for remote readings using a dial. There was also a demand for very precise thermometers. These were provided with auxiliary bulbs, and known as adjustable-range thermometers; the best of these were devised by E. Beckmann in 1888.

Gas- and vapour-pressure thermometers were an alternative to mercury instruments, but they are not often encountered; some of these experimental models had porcelain bulbs. In 1898 hydrogen was liquefied, and hydrogen was used in low-temperature gas thermometers; this type of thermometer succeeded in determining the melting point of oxygen in 1911. The late nineteenth century also saw a variety of bimetallic thermometers, in which different metals with unequal thermal expansion were welded together to form a strip. One end of the strip was clamped, the strip was formed usually into a U-shape, and the free end operated a pointer as it moved. Because of their convenient shape bimetallic thermometers made admirable pocket thermometers, and were enclosed in a case like a watch. A more unusual category of thermometer relied on the changes in the shape of a very thin metal tube filled with a highly expansive liquid such as ether, and it is interesting that one of the pioneers of aneroid barometers, Bourdon of Paris, was associated with these strain thermometers. A British patent was taken out in 1881.

A useful type of thermometer was the self-registering. High and low readings were retained for examination at a later date. Lord Charles Cavendish described what were called maximum and minimum thermometers in 1757, but the most important figure was James Six, who devised his instrument in 1782. This consisted of an upright U-tube, with a bulb at the end of each limb. The left-hand bulb and the upper part of its tube were filled with alcohol. The right-hand bulb was filled with alcohol, along with part of the associated tube. The rest of the tubing was filled with mercury; both ends of the mercury touched the alcohol. At these two points were very light steel markers held in place by delicate springs, just strong enough to stop the markers slipping down the tube when the level of the mercury moved in response to temperature changes. When the temperature went up, the alcohol in the left-hand tube expanded, pushing the mercury along until it pushed the spring-held marker up in the right tube. And vice versa for a fall in temperature. It was a simple and almost foolproof method. Steel markers were used for a very sensible reason; after each observation the markers were drawn back to the surface of the mercury by means of a magnet. Six's self-registering thermometers were also sold in a portable form, neatly packed in a wooden box complete with magnet. Pastorelli and Co and J. Newman were prominent makers of Six's thermometers.

Much the same principle was adopted in John Rutherford's self-registering thermometers of 1790 with regard to the steel marker, but Rutherford used two separate thermometers, mercury for the maximum temperature, alcohol for the minimum temperature. A more simple way of establishing a maximum temperature was that of the makers Pastorelli. A constriction near the bulb allowed the mercury to flow freely on expansion; this same constriction refused to permit the mercury to drop back into the bulb when the temperature fell.

Clinical thermometers are concerned, of course, only with registering the maximum. Introduced in 1866–7, the original clinical self-registers were 10in. long and bent, so that they could be read by the doctor when the bulb was beneath a patient's arm. It took five minutes to register the temperature. Not surprisingly this antediluvian apparatus was replaced a year later by the pocket clinical thermometer, which had a very fine bore. The thread of mercury was broken by an air bubble. When the temperature was going up the broken thread of mercury was pushed up before the bubble, but when the temperature went down the main thread of mercury dropped down towards the bulb end while the index section remained stationary. The index was reset by shaking the thermometer, exactly as in the modern clinical thermometer. The only snag was that a too energetic shake could cause the air bubble to lose itself in the bulb. The advantage of a very fine bore was that a widely-spaced scale could be obtained. Doctors in a hurry could make use of Pinketti's crescent clinical thermometer, in which the bulb was divided into two curved branches so that it would take the shape of the front of the bottom of the tongue.

More unusual self-registering thermometers included thermometers for determining underground temperatures (patented in 1874), a deep-sea thermometer devised about 1878, and bimetallic self-registers from the late 1860s. In the more esoteric range of self-registering thermometers the predominant makers were Negretti and Zambra.

Although the word pyrometer dates from 1749 it was then merely a synonym for thermometer, and it was not used in its present sense — a measurer of temperatures higher than those dealt with by a glass thermometer — for another century. The first pyrometer based on the phenomenon of the increase of electrical resistance of platinum with temperature was constructed by Siemens in 1871 in order to measure the intense temperatures of furnaces making steel. There is nothing attractive to a collector about this, as it consisted of a platinum wire coil wound on a core of firebrick and enclosed, with leads, in an iron container. The subsequent development of the pyrometer was hardly more exciting; a superior version of the Siemens pyrometer included wrapping the platinum wire round mica, and encasing the coil in a glazed porcelain tube which permitted heats up to 1200°C. Platinum resistance pyrometers were used for medical purposes, and detecting the heat developed by the beating of a frog's heart, a curious 1887 adventure.

Much more interesting and collectable than platinum resistance pyrometers are the earlier fumbling efforts to determine high temperatures out of the scope of the glass thermometer, such as the Wedgwood pyrometers, brass gauges with two tapering channels. Pieces of clay of a definite size were fitted into the wider part of one of the channels; after exposure to heat, these pieces were pushed along. How far they could go denoted the temperature to which the clay had been subjected. This principle was used in a pyrometer projected by J. F. Daniell in 1830. Byström's pyrometer, patented in Britain in 1863, consisted of a closed vessel containing a known amount of water with an outer case to prevent heat loss. At the top was a funnel mouth connected to a wire cage in the vessel; this cage could be rotated by a handle. A platinum ball was placed long enough in a furnace to reach the same temperature, transferred to the funnel, into the cage, where it was rotated as it cooled. The temperature of the water, ascertained by an ordinary thermometer, led to an assessment of the temperature of the furnace, the object of the exercise. Siemens also used a similar device, described by him in a technical journal of 1871.

In 1873 H. Hobson patented a hot-blast pyrometer to determine the temperature of air supplied to a blast furnace. This hot air was of the 700°C order, and naturally would destroy any ordinary thermometer. Hobson conceived the idea of introducing into his apparatus a known quantity of cold air. The thermometer could cope with the mixture of hot and cold air.

In dealing with the high temperatures of industry it was often necessary to take temperature readings a long way from the source, and dial-transmitting thermometers were introduced towards the end of the nineteenth century. Subsequently it was not difficult to replace the indicator with a revolving drum or disc, and although thermographs are interesting they are not really old enough to qualify as collectable instruments.

CHAPTER 10

PRECISION BALANCES

UNLIKE MANY scientific instruments where development was handicapped by there being no money to put into apparently senseless ventures, and which was carried out by enthusiasts and amateurs, sensitive and precise balances had been used for a long time by goldsmiths and money changers, and they only needed refining and improving. This was successfully done until by the nineteenth century accuracy to a millionth of the weight of the object weighed could be obtained simply by using physical apparatus consisting of an upright, a horizontal, and pans.

In the laboratory, assayers were the first to need accurate balances. Assaying, strictly speaking, is finding out how much metal there is in an ore, though it has a number of subsidiary definitions. Compared with later balances, those used by early laboratory technicians were clumsy. Assay balances consisted of a beam with a shaft passing through its centre, resting on bearings at the end of a vertical rod.

A typical balance of the early eighteenth century was that constructed by Francis Hauksbee in 1710, consisting of two arms of equal length from the ends of which pans were suspended. In the middle of the two arms was a swivel pin supported by bearings, and this was hooked on to a convenient

PRECISION BALANCES

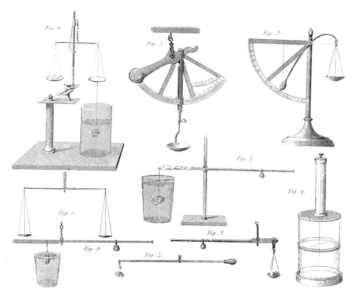

93 *A variety of balances, some tolerably accurate, others less so.*

horizontal. It was a suspended balance rather than one incorporating an upright, and derived from the balances of goldsmiths. Inefficient as a scientific instrument, the suspended balance had as its chief advantage its compactness. It could be folded up and taken about in a case with accessories such as tweezers and weights.

Improvements of the precision balance did not take place overnight. The pivots of the beam were better cut, and the bearings on which the pivots rested acquired a hard steel bushing. The wires of the pans were lengthened to lower the centre of gravity, and an indicator moving in front of a scale made it easier to obtain accurate readings. On later balances the scale was provided with a microscope. The pans were usually of brass and were suspended by means of three silk cords. An early improvement was to place the balance in a glass case, thus eliminating draughts and other outside forces that would deprive the observer of a true reading.

Between 1760 and 1770 important changes took place in the balance, and the first true precision balance was probably that constructed by John Harrison (1693–1776) for Henry Cavendish about 1770–5. Harrison guaranteed the rigidity of the beam by giving it a triangular section, and, realising that the key to the fine balance lay in the juncture of the horizontal and the vertical, he made the central knife-edge of the beam rest on top of a hard steel plate, doing away with bearings.

The knife-edge of the beam was extremely vulnerable, and it was necessary to lift the beam off the upright when the balance was not in use. Several methods were used to do this, including a system of pulleys, but the most practical was to insert a rod in a hollow vertical column. The top of the rod carried two curved arms with prongs which came out of slots in the upright and lifted the beam clear of the column. It was activated by a lever, which could be operated by a cord extending outside the case. Harrison's method was to use a metal plate fixed to a wooden beam inside the case. Two arms projected from the plate supporting the blocks on which the knife-edge rested, and the lifting apparatus extended into the base of the case, a cumbersome method which was not persisted with.

Harrison did away with the silk threads suspending the pans, and substituted thin brass wire, and the hooks on the beam were replaced by two cross links, which kept the horizontal movement of the pans to a minimum. The sensitivity of the balance could be regulated by threaded rods on the beam along which nuts could be screwed. One of these rods was placed vertically beneath one side of the beam, so that the centre of gravity of the beam could be transferred. The scale was no longer in the middle, but positioned vertically at the right-hand side.

The Harrison balance was much larger that previous balances used for precision work, and his pioneer work was built on by Jesse Ramsden about 1787, by Fortin in 1788 and by Mégnié. Ramsden used as his beam two topless cones joined at their bases to assure rigidity, and his knife-edge rested on agate plates supported by a brass frame consisting of four columns. The cone with the top taken off (known as a frustum) was a hallmark of Ramsden's, and was a brilliant device to obtain strength in his large theodolites—which were contemporary with his work on precision balances—and in the large refractor telescope he made for Sir George Shuckurgh in 1791. Topless cones acted as a much stronger tube.

Many of the great makers of mathematical and astronomical instruments were involved in the construction of precision balances, and Troughton also made a balance partly based on the one by Ramsden, with a single brass column replacing Ramsden's group of four. The balances made by Fortin in France had a big effect on later makers. Sensitivity was obtained by moving small weights along a threaded rod. It was important to have the precision balance exactly horizontal and although the obvious manner to do this was to use spirit-levels—Ramsden used two perpendicular to one another—Fortin preferred the plumb-line.

The balances of Ramsden, Troughton and the French instrument makers did not seem susceptible of much improvement, but about 1840–50 modifications were made, and precision balances assumed their modern form, with the brass beam in the form of a flattened elongated lozenge turned sideways.

It may be that the very simplicity of precision balances irked the more restless scientific minds, and about 1781–2 the Frenchman Lavoisier invented a pneumatic apparatus which he thought would have an application to balances. This device was known as a 'gasometer' and the idea probably arose from contact with steam-engine theory that was very much in the air at the time. The basis of the pneumatic balance is a tank inserted vertically into a slightly larger one, with nozzles and cocks fitted to let gases in or out. Variations in pressure and temperature were measured by thermometers and differential barometers built into the gasometer. The beam of the balance, resting on an orthodox vertical pillar, is terminated at both ends by two arcs of a circle, and from this is suspended, at one end, the pan, and at the other the pneumatic apparatus.

Although somewhat purposeless, the 'gasometers' represent an early involvement of engineers in measuring devices, an involvement that became much more important in the nineteenth century when all things were measured, from the amount of dew falling to the density of an indigo solution.

A substitute for the traditional precision balance was sought for in the balance operated by a spiral spring, but although useful in more mundane fields spring balances could never achieve the accuracy of the othodox precision balance. It was found at an early date that one spring tended to rub against the side of the tube enclosing it, and so customarily two springs are used, linked together, one going clockwise and one anti-clockwise. Absolute accuracy was also impossible because springs became stretched.

Precision balances are to be found quite easily, and even those of the early nineteenth century are by no means expensive. The smaller goldsmith's balances, complete with weights, and often beautifully encased, are also not uncommon.

CHAPTER 11

DRAWING INSTRUMENTS

DRAWING INSTRUMENTS were first used when the practice started of making designs and plans for constructional work preliminary to the work being done, and were developments of the tools, such as the straight edge, the square, and compasses, used by masons, carpenters, and other craftsmen involved in building. That the straight edge and compasses were used in antiquity is evident by a stele on an ancient Egyptian tomb of c4400 BC, by bronze tools found at Pompeii, buried by volcano AD 79, including compasses and calipers, and by items found on Romano-British sites.

Although most of the illustrations on the medieval illuminated manuscripts were done freehand, the borders were done with a stylus or a disc of lead in conjunction with a straight edge, and the introduction of paper about the twelfth century gave an enormous impetus to draughtsmanship. The preliminary work involved in designing the churches and cathedrals of the Middle Ages indicates that draughtsmanship was extremely sophisticated, though the first 'engineering' drawing dates from as late as 1573.

Draughtsmanship was handicapped by the reliance on pen and ink, and the use of graphite and the making of the pencil were first recorded in the 1560s. Early drawing instruments, except for the points, were usually made of brass, and occasionally of bronze, silver or gold. As soon as nickel silver was discovered draughtsmen preferred instruments made of this material, as brass gave off a slight odour. Nickel silver (brass whitened by the addition of nickel) was first made in Europe about 1824. Aluminium was tried, but found less suitable than nickel silver.

94 *A fine set of drawing instruments made during the George III period by a Dublin maker, and sold in 1972 for £290.*

Sets of drawing instruments made from gold and silver were highly prized. A case of instruments dated 1618 contained a pair of compasses with removable divider point, a stylus combined with a fluted pen, a crayon-holder combined with a gold pen nib, a parallel ruler connected by doubled links, a square folding like a knife, a 5in sector divided for linear and circular measurement, a pair of folding scissors, and a penknife.

Many instruments are much the same today as they always have been. The carpenter's square was modified into the T-square, and an early illustration of the T-square occurs in the French *Encyclopédie* of 1767. The taper blade of today came in some time between 1854 and 1866. The triangle, or set-square, followed the introduction of the T-square. They were first made from pearwood, but shrinkage across the grain made such set-squares inaccurate, and some time prior to 1850 pearwood or mahogany set-squares were edged with ebony and bevelled. In 1854 a patent was taken out to make set-squares of glass, and ten years later 'Goodyear vulcanite' was adopted, an unsuitable material as frictional electricity collected dirt and marked the paper. Celluloid proved the perfect material when it was invented and held the field until other plastics were developed.

The parallel ruler derives from the straight edge, and consists of two straight edges joined by links. There are two major types, the one where the straight edges are joined by two sloping parallel links, the other where the straight edges are joined by crossed or scissors links, the end pins of which slide into slots on the straight edge. Variations occur where there are more than two straight edges, such as the sectograph patented by Thomas Jones in 1811 which has upwards of ten straight edges. The rolling parallel ruler, first made by the instrument maker Dollond in 1770, consists of a straight edge combined with rollers on a rod.

The protractor is an instrument of sheet brass or other thin material, usually semi-circular, very occasionally circular, with a graduated circumference, and although the Oxford English Dictionary gives the date of its invention as 1658 the protractor was known before then. It was used by both draughtsmen and surveyors in the setting off of angles. As with the set-square, celluloid replaced brass and horn when it was introduced.

Until standard scales were introduced draughtsmen's instruments often had several scales engraved upon them, with their provenance noted. These various scales are often found on sectors, a sector being an instrument consisting of two straight edges hinged together. The word can have a variety of meanings, but the drawing office sector was invented by Thomas Hood and defined in his *The making and use of the Geometrical instrument called a sector* in 1598.

Until the arrival of graphite and therefore the ability to

make a marking material of precision the compasses were imperfect. Charcoal was no substitute for blacklead. With the coming of graphite the crayon-holder leg of compasses could be refined, and the sleeve replaced by a screw. This modification of compasses was first illustrated in 1569, less than ten years after the discovery of graphite. The first beam compasses were sketched by da Vinci in 1493, and were used for describing circles larger than those that could be drawn by ordinary compasses. In the beam compass the centre point and the tracing point are adjustable along the beam. Triangular compasses, one leg hinged at right angles to the other two, date from the seventeenth century, and are sometimes found in cases of drawing instruments of the period. Detachable points, certainly known at the beginning of the seventeenth century, made compasses more versatile, and tubular legs to compasses are credited to the engineer Brunel about 1799.

Bow compasses, first mentioned in 1796, are small compasses designed to describe very small circles, and are held by a head directly hinged to the points. Pocket or folding compasses had jointed legs turning inwards so that they could be made more compact for a case, and spring compasses and dividers substituted sprung metal for the orthodox joint. Dividers with screw adjustment were mentioned in Moxon's *Mechanick Dyalling* of 1703 for dividing quadrants, and spring dividers, with the spring adjusted by a screw, were known in 1738. Spring dividers are also known as hair dividers.

Proportional compasses are compasses with two pairs of points used in enlarging and reducing drawings, a task more easily undertaken by the pantograph, believed to have been invented by Christoph Scheiner in 1603. His pantograph consists of six rods jointed to form parallelograms, and was capable of producing a drawing of the same size, half the size or twice the size of the original. Later pantographs, such as that made by the Paris instrument maker Langlois in 1743, would give a three to eight times enlargement or reduction. In 1803 an American instrument maker patented a double pantograph, but there is no record of one being made.

Cases of draughtsmen's instruments were a stable bread-and-butter line for many of the most celebrated instrument makers, and, under a variety of headings such as cartographers' sets, can be found at reasonable prices. It became the custom to assemble drawing instruments in cases of leather about the middle of the sixteenth century. At first the instruments were carried vertically, and the cases sometimes include a sling so that they can be carried over the shoulder, but later they were placed horizontally in wooden cases with a velvet lining.

95 *Drawing instruments were a stable bread-and-butter line for many famous instrument makers, and this trade card of Tuttell illustrates a large number of such instruments.*

CHAPTER 12

SURGICAL INSTRUMENTS

SURGICAL INSTRUMENTS may be divided into two kinds, those in use before the discovery of antiseptics and those after. Earlier instruments were frequently ornamented and decorative, and only when it was realised that they were thus the breeding grounds of germs did surgical tools take on a purely functional appearance, and were preferably made from one piece of metal.

The instruments described and illustrated in the surgical writings of Ambroise Paré (1510–90) only altered in external details for hundreds of years, and only when it dawned on the medical profession that scrupulous cleanliness both in their persons and their instruments was the only way to keep patients alive was there a refashioning of such basic tools as scalpels and surgical saws. The scalpel of the sixteenth century folded into the handle like an ordinary pocket-knife, while the handle was adorned with a mythical or symbolic design, such as a winged female. Similarly surgical saws of the period were decorated and one saw had a richly chased metal frame, and at the end of the handle a lion's head with a ring through the mouth, to hang the saw up. Sets of surgical instruments often had saw and scalpel handles made from exotic materials such as bloodstone.

If one compares these instruments with those of 1850 the change is immense, but still the message relating to antiseptics

SURGICAL INSTRUMENTS

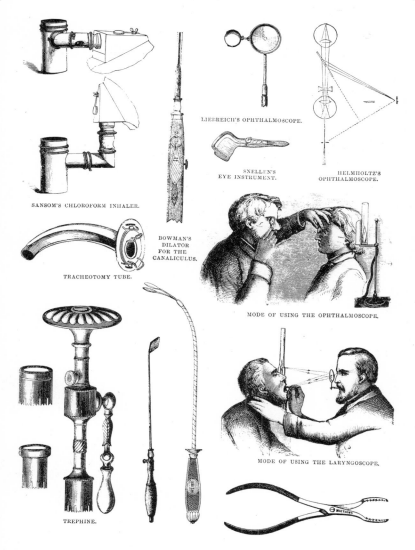

96 *A selection of early Victorian surgical instruments.*

had not fully got through, for the wooden or ivory handles of the mid-nineteenth century were etched with lines, like the lines

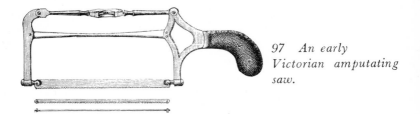

97 An early Victorian amputating saw.

on an engraving, and likewise harboured the dreaded germs. Wood or ivory as components of surgical instruments were shortly to be seen as not only dangerous but as inconvenient, for the boiling and immersion in carbolic lotion, standard precautions in the fight against germ warfare, played havoc with such materials. Instruments incorporating ivory and wood were replaced by all-metal tools, with handles nickel-plated. The aim of Victorian surgical instrument makers was to make their products smooth, simple and unrusting.

An instrument in which a degree of ornamentation remained was the needle-holder. Since the discovery of the ligature in the sixteenth century (arresting bleeding by tying up the blood vessels) surgical needles and needle-holders have varied a good deal. Older needles had slit-shaped eyes, not easily threaded, but in the nineteenth century needles were mostly flat with a round eye.

Other simple instruments that vary greatly in details are forceps and retractors. Catch-forceps were used to take hold of a bleeding point until it was ligatured, and towards the end of the nineteenth century a new form was brought in, known as the Spencer Wells forceps, with very narrow grooved blades as opposed to broad curved blades. Retractors are used to hold gently the edges of a wound while an operation is being undertaken, and they can be often recognised because they incorporate usually a broad, slightly concave polished surface for reflecting light into the wound. Retractors come in all shapes and sizes; some of them have the appearance of miniature back-scratchers, others bent toasting-forks, and others coal-shovels.

Scissors and probes and similar simple instruments follow the same pattern as forceps and surgical saws—the elimination of all ornament and inessentials, and the rejection of all materials but that which can be speedily rendered antiseptic. Without a machine-tool industry to provide identical parts, many of the earlier surgical instruments are somewhat crude, with abrasive working parts, but not until the MD replaced the surgeon–barber was anyone particularly worried about it.

98 Early surgical instruments are extremely rough and ready, as can be seen from this trade card of John Best.

The great advances in medical knowledge and techniques in the nineteenth century brought forth a host of new instruments, especially in the field of diagnosis, and although the basic tools of the operating surgeon remained the same, the Victorian doctor was able to make use of sophisticated instru-

SURGICAL INSTRUMENTS

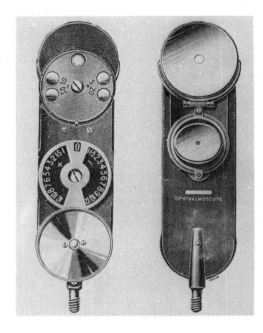

99 An early twentieth-century ophthalmoscope.

ments such as the ophthalmoscope and the laryngoscope.

Until the invention of the ophthalmoscope in 1857 by the great German physicist von Helmholtz (1821–94) it was impossible to study the interior of the eye. The laryngoscope, invented in 1860, made it possible for a doctor to inspect a patient's larynx by a combination of mirrors. The lithotrite was somewhat earlier. Although lithotomy, the operation cutting for stone in the bladder, had been carried out since the seventeenth century, the lithotrite was invented in 1839 for crushing bladder stones into minute particles so that they could be voided from the system.

Later inventors developed and sophisticated these instruments, and in view of nineteenth-century medical progress it is surprising that the tonsillotome, for taking out tonsils, did not arrive on the scene until 1881, though tonsillitis was known as a disease as long ago as 1801. Von Helmholtz himself, following on from his ophthalmoscope, evolved in 1864 the ophthalmometer, an instrument for measuring the curvature of the living eye by means of images reflected in it.

The use of anaesthetics revolutionised the surgeon's trade, and the more delicate and subtle instruments belong to the period after 1846 when the surgeon knew that he had time at his disposal to operate. Methods of administering chloroform brought their own instruments such as the inhaler, consisting of a mask, a valved container for the anaesthetic, and a rubber bulb operated by the anaesthetist. Ether was favoured rather than chloroform in the United States, and the administration of this was more tricky as ether had to be combined with the patient's own breath in proportions determined by the anaesthetist. The inhaler used for ether comprised a small metal chamber holding the liquid ether, a large india-rubber bag where the ether vapour was merged with the patient's breath, and a valved mask.

It is not surprising that electricity, a magic word to the nineteenth century, was brought into service as soon as possible, and late in the century the science of medical electrolysis was born. The passage of a small electric current between needles introduced into or under the skin brought about a shrinkage or cicatrisation of small growths without the consequence of an unsightly scar. Electrolysis is now used mainly for the elimination of facial hair, but the Victorian interest in the subject produced a number of quite interesting needle-holders, etc., associated with the topic.

Electricity was also responsible for an upsurge of interest in cautery, ie the arrest of bleeding using heat. Until the ligature was invented in the sixteenth century cautery was widely and painfully used; what was known as the *fer ardent* was the only way to stop a patient bleeding to death after amputation of a limb. It was still used to a limited extent even after the ligature had been introduced, and André Paquelin (born 1836) devised an ingenious form of cautery, using a series of metal blades or points fitted to a handle. These blades or points were hollowed out and filled with a fine platinum gauze. By means of a bottle and small bellows they were kept heated with benzine vapour, and when brought to a glowing heat by holding them over a spirit lamp they kept their heat whilst the operation was in progress. The electrical development of this was the use of platinum points kept at a red heat by means of a battery, and galvano-cautery sets may occasionally baffle seasoned collectors, much as electrical machines with an ostensibly therapeutic purpose (electrodes connected to batteries to induce a mild electric shock) may do.

Electric light brought forward a number of new instruments such as the cystoscope, which consisted of a long narrow tube, shaped and curved like a catheter, and having at the end what was described as a very minute glow-lamp and reflector, and a small window. At the other end was a lens, and a switch to operate the 'glow-lamp'. It was used to examine the interior of the bladder, and foreshadows in principle the use of tiny TV cameras in medical examination. A more sophisticated form of the cystoscope was the urethroscope, also using mirrors, lenses and light, and variations of the urethroscope were used in the examination of ear, nose and throat.

Surgical instruments are by their nature functional, and only early ones can be termed decorative. An exception might be made to the implements used by barber–surgeons for bleeding, which are usually found in attractive cases. Surgical instruments are also to be discovered nicely cased, and although their collection might not appeal to everyone the genre has interesting possibilities.

CHAPTER 13

MATERIALS AND MANUFACTURE OF SCIENTIFIC INSTRUMENTS

THE PRODUCTION of early scientific instruments depended on a suitable supply of raw material from which to make the instruments and of suitably skilled men to work the material. The manufacture of scientific instruments lagged behind their conception, and astronomers and scientists of all kinds were forced to kick their heels, waiting for a technical breakthrough that sometimes never came.

In the late sixteenth century the skill of metalworkers in southern Germany and the Netherlands in rolled brass led to a good deal of instrument making in those parts, and the availability of ivory in Dieppe, brought there by ship in the fourteenth century, led to an ivory-carving industry, so it is not surprising that this area became, centuries later, a centre for the manufacture of ivory sundials.

Ivory, bone, and wood were much more easily worked than metals, and in the early days, when cost was of no consequence, gold and silver were more easily manipulated than the harder but more suitable metals. There was also a good deal more prestige attached to gold and silver, and silvering and gilding were frequently applied to base metals when there was no logical reason. Silver was used in the seventeenth century for small sundials, or combined with ivory, and the precious metals are often in evidence in Arabic scientific instruments, such as the astrolabe, one of the few navigational aids used by mariners of the Middle Ages.

Of the woods, pearwood, other fruitwoods, and boxwood were especially favoured, and many navigational instruments later fashioned from metal were constructed from fruitwood. Instruments in which scales or measures were involved were obviously better in metal, but the engraving and incising of lines, numbers and complicated figures was difficult and expensive, and in navigation there were times when instruments had to be constructed without resort to instrument makers or other skilled men, as when navies and merchant fleets were being built up under the pressures of war or commerce.

It was much cheaper to make a master pattern in copper, and print from this copper plate the desired number of dials or calibrated scales on to paper, then stick this paper on to pasteboard or wood, which was then cut to the requisite shape. Many books of the sixteenth century specifically included large plates for this purpose. Boxwood was exceptional in that the scales could be engraved on the wood itself. Boxwood is of a very delicate yellow colour, dense in structure with a fine uniform grain. A reliable rather than a pretty wood, it was widely used by wood-engravers, and besides being used by instrument makers it was much employed by the makers of musical instruments such as flutes, since the time of the ancient Greeks.

Lignum vitae was used for such components of telescopes and microscopes as lens mounts, but the principal wood used in scientific instruments is mahogany, first noticed by Sir Walter Raleigh in 1595 but not used to any great extent until the early eighteenth century, when its suitability for furniture was observed by a cabinet-maker named Wollaston. By the 1750s more than 500,000ft of mahogany was being exported each year from its native Jamaica, and the demand became so great by the end of the century that in 1795 the tree was introduced to India. It remained the most desirable of wood until it was superseded in Regency and Victorian times by rosewood and walnut.

English instrument makers were very skilled in the manipulation of wood, and there are some telescopes of the early eighteenth century in which the tubes themselves are made from wood, but the favoured method of making tubes was by using cardboard covered with parchment, shagreen, or leather, though this way became obsolete with improved methods of dealing with brass. Cheap cardboard-tubed telescopes continued to be made. One of the last instruments almost totally made from wood was the 7ft Herschel reflecting telescope dating from about 1785, in which the problems of handling curved wooden sections were eliminated by having a hexagonal mahogany cross-section. In this instrument, the sighting telescope, the eyepiece tube, the mounts, and part of the rack and pinion elevating mechanism were made of brass. By the 1820s the entire telescope was being made of brass, including the base.

The use by British instrument makers of wood reflects general period design, using the natural grain, fine turning, and superb inlay. Islamic instrument makers of the eighteenth and nineteenth centuries had less respect for the natural qualities of wood, and quadrants, Quibla-indicators (forms of astrolabe), and other navigational and surveying aids made from wood were lacquered in bright colours, such as yellow, red and black.

So far as instrument making goes, the nineteenth century was the age of brass. In the early part of the century metalworkers preferred the red brasses to the yellow. Red brass contained 10–20 per cent zinc, while the yellow had up to 35 per cent. Yellow brass was harder to work than red brass, because the greater the percentage of zinc the harder the metal. Brass was worked cold, though experiments carried out in 1779 indicated that brass, even with a high proportion of zinc, could be worked either hot or cold. These experiments were capitalised on by the Birmingham brass manufacturer Muntz, who in 1832 took out a patent for an alloy containing 40 per cent zinc. In

1837 he started a factory in Swansea, which transferred to Birmingham in 1842.

It was found that Muntz metal, otherwise known as yellow metal, could be hot-rolled into sheets as easily as copper, and it was widely used throughout the century. German silver or nickel silver is brass whitened by the addition of nickel, and was first made in Europe about 1824, though it had no commercial impact until 1840 with the introduction of electroplating. Zinc was also used in pinchbeck and was the main ingredient of spelter, but neither of these alloys played much part in instrument making.

Nickel itself was first worked about 1880 after it was found that it could be toughened by adding a little magnesium, and copper–nickel alloys appeared on the scene shortly afterwards. Aluminium was occasionally used by the Victorians, but not so often in instrument making as one would have thought. Isolated in 1754, the metal was obtained in 1827 but it was not until 1856 that it was a commercially viable substance. Aluminium, of course, is very light, malleable, does not rust and is impervious to all acids except hydrochloric. In March 1856 it was £3 per ounce, but by the following year it was down to 11s and gradually became yet cheaper. Prominent in putting aluminium before the public was H. St-Claire Deville, who in 1859 wrote *De l'Aluminium,* and the prospects sufficiently intrigued the French government for them to make the eagles of the French colours from aluminium.

Newcastle-on-Tyne became the centre of the British aluminium industry, and in 1862 a new alloy of copper and aluminium was evolved, called aluminium-bronze, and this is occasionally met with, though not to the extent that the inventors would have imagined. Aluminium-bronze was used especially for watch cases. Aluminium itself enjoyed an extraordinary vogue in the 1890s, and dropped in price from £1 a pound in 1889 to 1s 7d in 1895. It arrived too late in the century to play a major part in the instrument-making industry, which, in contrast to the seeking for novelty at all costs that marked Victorian progress, was conservative and cautious, and was more concerned with the assimilation of photography and electricity into the field of scientific instruments than with the unquestionable advantages of aluminium.

Phosphor bronze was perhaps of more significance. This alloy was first heard of in 1871, and it contained 0·1 per cent phosphorus and up to 5 per cent of tin. The tin bronzes had been known for upwards of 2,000 years, and had been used for the making of statues, but they were difficult to work. From the mid-1870s phosphor bronze was increasingly used in electrical apparatus, its high tensile strength making it extremely suitable for springs and electrical contacts.

None of these alloys was considered an adequate substitute for brass, a material that was also preferred to iron and steel. Most scientific instruments involve tubing, and metalworkers preferred to work in brass. Until the nineteenth century, tubes were made from metal rolled out into a narrow strip; this was folded lengthwise over a mandrel bar, leaving an overlap, which was welded while hot. A new method was used in the early part of the nineteenth century, involving the casting of brass tubing around a central matrix, and a further method perfected by Kynoch in the manufacture of brass cartridge cases, known as cupping and drawing, was adopted for the production of small thin-walled tubes.

Silver plating was developed as a cheap alternative to silver. Old Sheffield plate was first made in 1743 — copper with a coating of sterling silver on one or both sides. It was a cumbersome process and relatively expensive, and in 1840 it was dramatically replaced by electro-plating. A patent taken out by the Elkingtons of Birmingham proposed the use of the electric current. Articles required to be plated were placed in a bath of potassium or sodium cyanide solution in which gold or silver, or their oxides, had been dissolved. Current was supplied through a bar of zinc or other suitable metal. Without the complications of Old Sheffield plating, copper, brass and German silver were miraculously coated with a thin layer of gold or silver. This method was eminently suitable for objects of an intricate or tortuous shape, especially those associated with the home such as teapots or urns, or with functional things like cutlery, and until 1890 Britain enjoyed a virtual monopoly, though she was later overtaken by Germany.

To the instrument makers silver plate was seen as a cheap substitute for the real thing, and the staid makers of the old school preferred to stick to brass uncluttered by a silver coating. Steel was used a good deal, but not to the extent one would have thought, though improved methods of processing it (by Bessemer in 1860, by Siemens in 1876) did not go unnoticed. Iron was much coarser than brass to work, though it was widely used in its cast form for bases, and in sheet form as the barrels for astronomical telescopes such as the 20in. reflector made by James Nasmyth in 1842.

The role of the instrument maker changed less in the aftermath of the industrial revolution than it did in other trades. The instrument makers remained a clique, an élite, who to some extent responded to the great demand for their wares but who preserved their integrity intact. The workshop rather than the factory remained the social background throughout the nineteenth century, and there was a good deal of interaction between the instrument makers, one man specialising in microscope mounts, another in the production of microscope stages, another skilled in tripod stands, while the division between the mechanics and the optics of many scientific instruments meant that few instrument makers were able to make a telescope or a microscope entirely in their own workshops.

Many of the makers were family concerns, and apprentices, when they set up on their own, often moved a few doors away and continued to work in close alliance with their former employers. Occasionally friendly rivalry between the instrument makers would harden, as we can see in the plethora of law suits that followed the breaking of the Dollond achromatic lens patent.

The financial revolutions that transfigured British industry did not make much of an impact on the instrument makers, who were backed by rich patrons and who did much of their best work for aristocratic dilettantes. Parliamentary bills in 1844 and 1856 made it advantageous for many trades and industries to go public, and in the 1860s a growing number of firms took advantage of the laws to raise from the public the money they needed to expand and improve their industries.

Instrument makers were influenced less by economic than by technical conditions, and no matter how skilful were the individuals their products were conditioned by the tools they were working with. Machine-tools made it possible to shape and fashion metals with an accuracy that was impossible by hand.

Until the eighteenth century tools were simple and unsophisticated, and those that were used derived from the clock-making industry. Robert Hooke, author of *Micrographia* (1665) and one of the greatest figures in microscopy, invented in 1671 a machine for cutting clock wheels which used a rotary file to cut the teeth, which were then smoothed and shaped by hand. In 1725 the clock-maker Fardoil made a file-cutting machine, in 1751 Nicolas Focq constructed a mechanical planing machine, but more important than either Fardoil or Focq was Jacques de Vaucanson. Vaucanson is better known as the maker of marvellous automata, but in 1760 he invented a slide lathe fitted with a tool-holder which could be mechanically operated.

The lathe is the most important instrument of them all, defined by the OED as a machine for turning wood, metal, ivory, etc, in which the article to be turned is held in a horizontal position by means of adjustable centres and rotated against cutting tools. The lathe itself goes back into antiquity, and by the seventeenth century it was widely used for the shaping of wood and ivory, and, more rarely, iron. The lathe of the eighteenth century incorporated two novel features — it could be mechanically operated, and it had a tool-holder. In older lathes the tool was held by hand, and supported by a block parallel to the axis of the lathe. Nevertheless, such lathes were able to produce a variety of ornamental and decorative work by eccentrically moving mandrels, and we see the results in the wooden parts of microscopes by Mann (1660–1730) and Marshall (1663–1725).

The development of lathes was helped by the enthusiasm of rich amateurs with their own workshops, and in France Louis XVI himself was a keen manipulator of the lathe. The lathe was invaluable to instrument makers, and in his book *L'art du tourneur méchanicien* (1775) M. Hulot declared 'It is not from the turner's skill that the makers of instruments for mathematics, astronomy, physics and hydraulics are indebted for their essential needs? Of what use would these instruments be without the exactitude which only the lathe can achieve?'

Lathes improved tremendously during the eighteenth century, and whereas in the past they had been made wholly of wood, now they were constructed of iron, steel and brass. More significant than Vaucanson in the development of the lathe was Henry Maudslay, the most gifted of eighteenth-century toolmakers. Many of his inventions arose from his employment with Joseph Bramah, a genius who invented a wood-planing machine, a device for numbering bank-notes, an improved water-closet, and, above all, an all but impregnable lock. In 1784 he patented a lock and offered 200 guineas to the man who could pick it, a wager that was not won until 1851 when a mechanic did the trick, taking 51 hours over it.

Maudslay (1771–1831) was as yet unborn when Vaucanson made his improvements to the lathe, but before the close of the century he had left Bramah and set up on his own. One of Maudslay's most important innovations was in the production of accurate screws, vital in the manufacture of such working parts as the slow motion of microscopes. Maudslay made a wooden screw, then copied it on the lathe in metal. There was a spin-off in the machine-tool industry from the precision with which he turned screws, for the accurately threaded screw was a key part of a bench micrometer he made, accurate to 0·0001 in. He was an early exponent of standardisation, the theory of replaceable parts which led eventually to the techniques of mass production.

Like instrument making, machine-tool making was self-propagating. Just as Maudslay had worked for Braham before branching out in his own workshops, so did Maudslay employ men who were later to prove eminent in the field, even surpassing him in some respects. Richard Roberts made in 1817 a lathe of an extremely advanced design, though most of his innovations were of more use in the manufacture of large items, such as a machine for punching holes at regular intervals in steel plate. Another of Maudslay's protégés was James Nasmyth, inventor of the steam hammer and a keen amateur astronomer who constructed his own telescopes. In 1829 Nasmyth made a milling machine of advanced design, cooled by water dripping on to the work being machined, and in 1836 invented a shaper for planing small surfaces and producing any surface dominated by straight lines.

Important as Roberts and Nasmyth were, they were eclipsed by Joseph Whitworth, who dominated mid-Victorian machine-tool technology with a wealth of inventions and improvements. In the Great Exhibition of 1851 he had 23 exhibits covering the whole field of machine tools, many of which were used in instrument making and caused a complete reappraisal of the industry. So admirable was his workmanship that a lathe made in 1843 was used until 1951. Whitworth reinforced the work done on the standardisation of screw threads by his one-time master, Maudslay, and in 1841 he proposed a constant angle (55°) between the sides of the threads, and a set number of threads to the inch for the various diameters. Precision and accurate measurement were his guide-lines, and as early as 1834 he made a measuring machine that operated to a millionth of an inch, an incredible achievement for an age that was only just coming out of the period of clockwork.

Machines for making machines were jerked abruptly into a new dimension, and this affected instrument making. No longer were the makers necessarily craftsmen, individuals, egocentrics, left-overs from the old Guilds of the Middle Ages. The mysteries of their crafts were being dissipated. It took a long time for the message to get through; one doubts if it ever really got through in Britain, though unquestionably it did in America, which seized the new techniques eagerly, got mass production started, and branched off into technologies that made Britain appear to be living in the stone age. It was no accident of history that nineteenth-century marvels such as the sewing machine and the typewriter began life in America, products of the modern machine-tool and its accompaniment — identical components and production-flow techniques.

A great innovation that owed little to Maudslay and his successors was the turret lathe, the name given to a multi-purpose lathe in which a rotating turret could carry as many as eight cutting tools. An operator could carry through a complete process, turning from one tool to another without moving from his machine. Unskilled workmen could use these, and the most important person on the factory floor became the toolmaker, who adjusted and replaced those cutting tools that needed it. The turret lathe was extensively developed in America, where its genesis lay. There was a shortage of labour in America, and any method that would short-circuit this was welcomed.

Many of the inventions of the nineteenth century passed the instrument makers by, though the machine-tool revolution made possible, for example, the first telescope factory rather than workshop in 1855. Situated far from London, this factory had to rely on relatively unskilled labour, and would have been unable to function but for efficient machines.

The London instrument makers carried on as they had done for a century or more. The prestige makers did not need to modernise their operation — the fact that their capacity was unequal to the demand was of only marginal consequence. The one modern instrument that the instrument makers accepted without reservation was the dividing engine for producing accurate and repeatable scales, necessary in most surveying and navigational instruments. The vertical and horizontal circles of theodolites, the Vernier scales, the great mural circles used by

MATERIALS AND MANUFACTURE OF SCIENTIFIC INSTRUMENTS

astronomers, all these were precision instruments and components demanding to be precision made.

The eighteenth century used instruments that checked on instruments that made instruments, such as the diatometer, for measuring the thermal expansion of metals, or comparators (a beam compass the points of which are moved by micrometer screws). The nineteenth century did not need them.

Though probably less romantic than eighteenth-century instruments, there is no question that nineteenth-century scientific instruments are more efficient in every way. And there are many more of them, thanks to the machine-tool industry.

CHAPTER 14

CLEANING AND RENOVATION OF SCIENTIFIC INSTRUMENTS

It is still not unusual to find the more out of the way scientific instruments tucked away in the corners of junk shops, the owners of which have no idea what the objects are, and although this is not true of instantly recognisable things such as telescopes and microscopes, even these can be found in a sorry state. The collector on finding an instrument in need of a good refurbishing will have a choice of whether to bring the instrument up into tip-top condition, with gleaming brass and immaculate woodwork, or to merely remove the dirt, grease and rust and have an interesting *objet d'art* rather than a working instrument. Some collectors even think there is something vulgar about shining brass.

The author cannot share these delusions. A beautiful object should be presented at its best, whether it is a picture, a piece of furniture, or an instrument. Scientific instruments, unless one has an exceptionally deep pocket, are unlikely to qualify as objects where the main attraction is age, as in old oak, where the machinations of long-dead woodworm and the inroads of the occasional death-watch beetle add to the aesthetic appeal.

A scientific instrument can be divided into a maximum of four parts: (a) the metalwork, (b) the woodwork, (c) the optics (if any), (d) the covering (if any). By covering is meant the parchment, leather, shagreen, etc., used to cover the tubes of microscopes and telescopes.

The metal will more often than not be brass, and brass is very responsive to attention. If the metalwork on a scientific instrument is in a bad way, covered with grime, grease and a generation of dirt, one should not have preconceived ideas on what it should eventually look like. In the early nineteenth century workers in brass preferred the red brasses, as they were easier to work, but with the patented method of Muntz brasses became much yellower. Muntz metal, or yellow metal as it is often called, should be treated as brass.

It is advisable to strip down the instrument as much as possible, though care should be taken with parts where there are fine adjustments. The tubes of microscopes can be removed from their mounts, and it is very important either to remove the optics or to protect them if removal is impossible. Quality instruments usually have detachable eyepieces and objectives, and any interior lenses can be protected by stuffing the tubes with cotton wool.

Unless one is positively brutal, brass will take a lot of hard treatment, and one should not approach the job of cleaning it too hesitantly. The first step is to use a nail-brush dipped in petrol to remove all traces of old grease and grime. When this is done, metal polish should be applied liberally with another nail-brush, left for a few minutes, then rubbed vigorously in a lengthwise direction. This should be carried out until one is certain that beneath the dirty liquid the brass is clean; this can be checked by rubbing a small portion clear.

The metal polish should be washed off with medium-hot water to which has been added detergent, but care should be taken not to let the water get into the tubes of microscopes or telescopes. The next step is rinsing and drying the brass, using an old towel, and the final polishing is done with a little silver-plate powder mixed with methylated spirit, applied with a clean brush. This is then brushed off with a soft clothes-brush, an operation better carried out in the open air as the dust hangs about. A clothes-brush, although ideal for large areas such as the tubes of telescopes, may be too cumbersome for smaller pieces, in which case watercolour brushes can be used. For small items it may be that a nail-brush is also too unwieldy, and an old tooth-brush, preferably bristle, not nylon, substituted.

For a more cursory cleaning impregnated cotton wool, such as 'Duraglit' may be used, but there is nothing like the full treatment for imparting to brass the sheen it deserves. If there are any obstinate discolorations these can be removed during the cleaning process by wire-wool dipped in metal polish.

One of the problems in renovating brass is getting rid of old lacquer and applying a new coat. There is some doubt about the precise recipe used by the old instrument makers, though the basis of their lacquer was certainly shellac. A nineteenth-century recipe for getting rid of old lacquer can hardly be improved upon even in these days.

A quantity of clean wood ashes is boiled in water, and soft soap is added. The brass article is immersed in this, and the lacquer quickly disappears. The work is then dipped for five seconds in a solution of aquafortis and water, sufficiently strong to clean off the dirt, then washed immediately in clean water. The brass is dried in sawdust, and rubbed with a chamois leather, then heated to about $210°F$, perhaps on a hot plate or in an oven. It is now ready for the application of laquer.

Obviously this approach is too cavalier for brass components that for example contain interior lenses that are difficult to remove, in which case methylated spirits can be used to remove old lacquer and the tube prepared for its new coat by being heated in the steam of a kettle. A very good cold lacquer for brass can be made by dissolving 2oz of ground turmeric, 3 drams of gamboge, 7oz of powdered gum sandarach, and $1\frac{1}{2}$oz of shellac in 2 pints of spirit of wine. This mixture is shaken, dissolved, strained, and augmented by $\frac{1}{8}$ pint of turpentine varnish.

It may be thought desirable to use the same colour lacquer as was used originally. For a yellow as opposed to gold lacquer,

2oz of shellac, ½oz of gamboge, and 2oz of turmeric are dissolved in 2 pints of methylated spirit. Red lacquer is made by mixing 2oz of annatto, 1oz of shellac, and 1oz of dragon's blood in 1 pint of methylated spirit. Blue lacquer is obtained by grinding Prussian blue in pale shellac varnish (white shellac).

One of the most closely guarded secrets of those few dealers who actually work on their acquisitions is the ageing of brass. But secrets are made to be broken. Presented in chemical terms the recipe is Ba_2S_3 (½oz) plus K_2S (⅛oz), or in common parlance ½oz of barium sulphide and an ⅛oz of potassium sulphide. This is perhaps the most useful of the recipes, for it enables the operator to match up new treated brass components with the old brass.

Modern renovations and additions can also be concealed (if that is desired) by blackening the brass. A strong solution of washing soda is added to a solution of copper sulphate (bluestone). This combination is allowed to settle, the liquid is poured off, and a quantity of water equal to the liquid poured off is added. This is allowed to settle again, and as much liquid as possible is poured off once more. One is left with a green sediment, and four times its quantity in water is added, heated to 140°F, and this provides the bath for the brass. Ammonia is added gradually until the required colour is obtained.

The classic way to clean copper and brass is a 'pickle' composed of two parts of sulphuric acid, one of nitric acid, and two of water. Many amateurs fight shy of dealing with such a solution, but one is safe if one remembers always to add *acid to water,* never the other way about, wear rubber gloves *always,* and protect the eyes and the face from splashes by wearing goggles or a mask. When the copper or brass is cleaned by immersion in this solution, it is washed in cold water.

There are few problems in lacquering. It is a job that anyone can do. It is laid on evenly with a camel-hair brush, care being taken that the brass is not gone over twice. The brush should be stroked only one way, not up or down as in painting, or the lacquer will appear uneven and streaky. The solvent used for preserving brushes is methylated spirit.

Gilding need present no more difficulty than lacquering. For brass and bronze, $\frac{1}{4}$oz of gold chloride is dissolved in 5 quarts of distilled water. To this are added $2\frac{1}{2}$lb of caustic potash, 5oz of pearlash, and 2oz of cyanide of potassium, and the solution is brought up nearly to the boil, whereupon the articles to be gilded are dipped. The colour of the gilding is marginally affected by the temperature, and trial and error methods may be needed to obtain the most agreeable tone.

Prestige scientific instruments were sometimes silver gilt, and are therefore by definition valuable pieces, and it would be understandable if owners of these would prefer to have regilding done professionally. However, there are no real problems attached to gilding silver or German silver. German or nickel silver is not silver at all, but an alloy of copper, nickel and zinc. It is harder than silver, but has an unmistakable grey tinge. Exposed to the air it turns slightly yellow, and often needs treatment to prevent it from looking like anaemic brass.

A gilding liquid for dipping silver and German silver is obtained by dissolving 20g of gold chloride in a pint of distilled water, then adding gradually $1\frac{1}{2}$oz of acid carbonate of potassium. This is mixed with another solution of $1\frac{1}{2}$oz of acid carbonate of potassium and 1 quart of water. The entirety is then boiled until it turns green. Gilding by use of this liquid may be made more permanent by first applying a thin coating of mercury in a solution of nitrate of mercury.

Cast-iron bases can be dealt with in a more summary manner, by washing in warm water and detergent and briskly going over with a wire brush. In steel or iron components, rust is a problem, and it is a good idea when buying a really derelict instrument to ascertain how deeply the rust has bitten in. Rust usually looks much worse than it really is, and any but the most virulent attack can be countered by the kind of rust-remover sold by garages. If the instrument is in itself desirable, a certain amount of pitting should not put the purchaser off; it is the rust that looks bad, not the pitting, and if the steel is brightly polished even a considerable dose of pitting will be overlooked.

For large instruments in which the rust has eaten through and made holes resort can be made to fibre-glass or plastic metal. Although plastic metal is rarely as good as the makers maintain, and is clearly not metal, it unquestionably has its uses. Bronze is not so often found in scientific instruments. It is made of copper and tin, with occasionally a little zinc and lead, and can be cleaned in much the same way as brass, though it is more vulnerable, and the patina on old bronze caused by oxidisation may be so aesthetically pleasing to the collector that he is tempted to eschew a thorough cleaning. An interesting and almost unknown way to emulate bronze is to coat a light metal, such as spelter, with brown poster paint and rub on brown bootpolish. This can be experimented with if the patina is too objectionable.

Pewter and aluminium are less often found in scientific instruments than brass or steel, but they can both be cleaned by mixing powdered pumice with chalk, and washing with water and soda. Marble may be encountered in one-offs, and if this is stained it can be treated with whiting powder and white spirit, brushed on and left overnight. The resultant powder can be taken off the following morning with a fine hair brush. One frequently finds ivory on scientific instruments, used as slides for microscopes, on microscope accessory drawers, for makers' names, or simply as ornament. Scratched ivory can be polished with fine pumice powder and petrol, and brought to a pristine state with whiting powder and methylated spirit applied with a felt pad. Yellowed ivory can be bleached by exposing to sunlight. Late instruments not in the top echelons as regarding quality may contain simulated ivory — celluloid, xylonite, or other compositions. These betray themselves by giving out a smell of camphor when wiped with methylated spirit, but they can be cleaned by using pumice powder and petrol, as with ivory.

The woodwork of scientific instruments is either in the solid or in the form of veneer. The wood most used in the solid were fruit woods and boxwood. The surface finish was often beeswax, and to remove the accumulated dirt of centuries it is necessary to remove the beeswax. The best way to do this is by using a proprietary brand of paint stripper, but the collector must judge for himself whether the instrument is too valuable to subject it to such treatment. An old wooden sextant might well cause one to have second thoughts. But in the case of bases of microscopes or the legs of theodolites and other surveying instruments there need be less hesitation, though advice should be taken where mahogany is concerned, one of the most vulnerable of all woods to cavalier treatment.

The stripper should be applied with a paint-brush, one surface at a time, and when this process is completed methylated spirit should be applied to the wood to counter the caustic effect of the stripper. When rewaxing, a furniture polish using a basis of beeswax (such as 'Antiquax') should be used, and not one with a silicone base. For the true enthusiasts a polishing wax as prescribed by Sheraton himself can be used:

> Take beeswax and a small quantity of turpentine in a clean earthen pan, and set it over a fire till the wax unites with the turpentine, which it will do with constant stirring about; add to this a little red lead finely ground upon a stone, together with a small portion of fine Oxford ochre, to bring the whole to the colour of brisk mahogany.

The fine microscopes of the eighteenth and early nineteenth centuries frequently incorporated veneers on their bases. These bases often had drawers for the accessories, and were the products of the cabinet maker rather than the instrument maker. Lifting veneer can be a problem; if a small portion of veneer is lifting, it probably means that the adjacent bits are loose, in which case it is better to take it all up and reglue the whole section. The modern amateur has a wide choice of glues at his disposal, and no doubt everyone has a preference for one

CLEANING AND RENOVATION OF SCIENTIFIC INSTRUMENTS

over another. One of the most reliable is 'Evostik' woodworking adhesive, a white substance that is refreshingly unmessy.

The bubbling up of a veneer might seem to present an insoluble problem, but there is a quick and ingenious way to deal with this. A hypodermic needle, available from any large chemist, is filled with glue and the point inserted into the bubble. The needle is pushed home until this space is filled with glue, and when the needle is withdrawn the excess glue will escape through the hole when the veneer is pressed down and weight applied. Depending on the size of the needle, a thinner glue than 'Evostik' such as 'Seccotine' may be needed.

Veneers were French polished, and, like beeswax, a good deal of dirt could be locked in. On no account should paint-stripper be used on veneer, and the only way to remove French polish is to use a very fine sand-paper. Today there are many short cuts to French polishing, and a gloss can be put on the veneers without too much expenditure of time.

French-polished surfaces can be cleaned with a mixture of half a teacupful of common washing soda to a gallon of water, and revived by using a combination of lime water, raw linseed oil, and turpentine in equal parts. The water and oil are well shaken until thoroughly mixed, then thinned with the turpentine. This is then applied liberally to the surface of the wood by means of wadding, rubbed well, and cleaned off with a cloth. Another piece of soft cloth is sprinkled with methylated spirit, and pressed against the wooden face until it presents a moist (not wet) surface. This cloth is then used to rub the surface, and a second pad is taken richer in methylated spirits to clear away any trace of oil. Friction then imparts the gloss. Another reviver, vinegar, linseed oil, and methylated spirit, is used in the same way.

Repairs to woodwork should be carried out before cleaning or polishing. Joints can be reglued, and if there are no clamps available a make-shift clamp can be made by using soft rope and a piece of wood, applied like a tourniquet.

Veneer can be taken off quite easily by using an old-fashioned flat-iron. The heat melts the glue, and the pieces of veneer can be lifted off using a flat knife. Another method is to cover the surface with a pad of newspapers and pour boiling water, a method that will also melt the glue.

Obviously old-fashioned flat-irons and pads of newspaper are useless when the area of veneer is very small, as in the facing of a microscope accessories drawer, and a method that can be used here is to make a ring or 'wall' of putty round the veneer to be removed. Linseed oil is then heated in a water bath until the water boils, poured over the veneer and left for about three hours, by which time the oil should have soaked through the veneer and softened the glue. A thin-bladed knife, such as the common table-knife, will then be used to lift the veneer.

Woodworm can be disfiguring though it is not the terror it once was, and useful composition for filling in woodworm holes is whiting powder and linseed oil made into a thin paste, and coloured to match the wood. At the moment of using the mixture a little French polish should be poured in and mixed well — this hardens the composition.

Books on furniture restoration speak airily of going to your veneer stockist for veneers. Unfortunately one can travel hundreds of miles without finding anyone stocking veneers, and antique dealers often buy nondescript pieces of furniture just for the purpose of obtaining veneer. One antique dealer in Hastings habitually has his bath full of floating pieces of wood, and when he takes a bath he has to fight his way through the flotsam and jetsam. An easier way of obtaining veneers is to buy a marquetry set, sold in most craft shops, which usually contains a variety of veneers.

The pieces of veneer are cut to shape with those knives with interchangeable blades sold by craft shops and ironmongers, and when they are glued and set in place it is a good idea to cover it all with adhesive paper, which can be soaked off when the glue is dry. It is no difficult task to make a matrix and cut veneer shapes, but more problems may be met with inlay stringing, necessitating cutting lengths of inlay perhaps a sixteenth of an inch wide. This is very much a trial and error operation, though basically all one needs is a sharp blade, a metal straight-edge, and a good eye. If possible one should make the inlay stringing of metal or white wood, which has a closer texture and is therefore less brittle than some of the darker woods.

Although the wooden parts of scientific instruments are usually less important than the metal — except in early nautical instruments — a damaged base beyond repair or a microscope accessory box in fragments naturally detract from the appearance and the value of a piece, and often it is necessary to have a new section made up. In this case it is logical to age the wood to correspond with the metal. This can be done in several ways, one of which is fumigation. Fumed oak was a favourite of the Victorians, and the process can be extended to other woods including mahogany.

Fumigation involved placing the article in an air-tight cupboard or box (such as a packing-case or tea-chest) on the bottom which are dishes of liquid ammonia ($\frac{1}{4}$ pint is sufficient for 80 cu ft). Fumigation enriches and darkens. Oak can be processed in shades from olive to dark brown. Certain types of oak and other woods will not react to ammonia, and it is a good idea to try the piece out by taking out the stopper of the ammonia bottle and placing the bottle mouth against the wood to see if the wood changes colour.

Although not so efficient, wood can be aged by using bichromate of potash in a solution of just 1oz of potash to a pint of water, applied three or four times. Various colour effects can be obtained by using 'red oil' (2oz of alkanet root in $\frac{1}{2}$ pint of linseed oil) washing soda, water slaked with lime, and carbonate of soda. A French method is to rub the surface of the wood with dilute nitric acid which, when dry, is brushed over with a solution of $1\frac{1}{2}$oz of carbonate of soda dissolved in a pint of methylated spirit. The great advantage of fumigation is that it will not raise the grain.

The coverings, such as leather or shagreen, can be refurbished using a preparation such as Probert's leather polish, used by saddlers and the antiquarian book trade. If the leather has suffered from acid decay, a solution of $7\frac{1}{2}$ per cent potassium lactate in water should be applied before the leather dressing. Leather of the kind suited for the covering of telescope tubes can be quite easily obtained; cabinet-makers usually have some in stock, for 'antique leather' is widely used in the renovation of 'partners' desks' and similar highly sought-after pieces of furniture. Glue is used to fix the leather to the tubes.

The old method of cementing leather to metal was by using marine glue, soaked in cold water until soft, then dissolved in vinegar at a gentle heat, then supplemented with a third of its volume of turpentine, mixed, and applied hot. The metal was usually prepared by being painted with white-lead and lamp black. Marine glue is made of one part indiarubber, twelve parts of naphtha, and twenty parts of powdered shellac.

The leather coverings of telescopes and microscopes were often gilded with decorative devices, and although experience in bookbinding is useful, the layman can renovate such coverings. Gum mastic in fine powder is dusted over the surface, and an iron or brass tool bearing the design to be impressed is heated and gently pressed on a piece of gold leaf, which adheres to the face of the tool. When the tool is pressed against the leather, the mastic softens and retains the gold, and all superfluous mastic is brushed off. The design is then lightly dabbed. The same method can be used with silver leaf.

If the leather covering on a telescope is far gone it may be better to strip it off completely than attempt to renovate, and a good method of doing this is to apply a strong solution of caustic soda.

Shagreen originally came from the hides of the wild asses of Persia and Turkey. Its surface was rough with small round lumps, and it was used in lieu of sandpaper as well as being an alternative to leather. Natural shagreen was grey or white, but for the coverings of scientific instruments it was usually dyed.

The favourite colour was green, made from sal ammoniac and copper filings, but it was also dyed blue with cobalt, red from cochineal, or black. So great was the demand that an ersatz shagreen was evolved, from the hides of horses and camels, with the lumps artificially put in. Small seeds were sprinkled on the flesh side of the 'shagreen' when wet, pressed in, and beaten out when they had done their job.

In the eighteenth century shagreen took on a new meaning. Shagreen was now the skin of sharks. Like true shagreen it took green better than any other colour, and by the end of the eighteenth century animal shagreen was little used. The granular surface of shark-skin made shagreen very hard wearing and, being waterproof, it was admirable for hand telescopes that came in for rough treatment. The surface very quickly got scruffy and dirty, but it responds well to a thoroughly good clean. Instrument cases can often be found covered with shagreen. The present centre of the shagreen trade is Biarritz, and for purists wishing to replace shagreen coverings it is still possible to obtain the material. Coverings of parchment and vellum were less often used. Parchment can still be obtained. The cheapest way is probably to buy old deeds and documents, often seen on sale for a pound or two in antique shops, and remove the writing. By soaking the parchment it becomes flexible and amenable.

Among the most vulnerable parts of a microscope is the mirror located beneath the stage to direct light into the lens. A cracked mirror will not necessarily detract from the efficiency of a microscope but it looks bad. Fortunately dentists' mirrors are ideal as replacements, and although they are 1/16in smaller in diameter than the standard microscope mirror this margin can be eliminated by gluing a circle of string round the edge of the mount. Some difficulty may be found in removing a broken mirror from the mount, for it is kept in place by a slightly overlapping rim of thin metal. It is easy enough, of course, to be brutal and wrench the mirror from its backing, but this means that the rim will be broken or twisted and will not be of any use in holding tight the new mirror. There is a simple way out of this dilemma; take a hack-saw and carefully cut across the surface of a broken mirror and the rim. The rim will thus be neatly cut (the recommended number of cuts is about eight), and each piece can be lifted up, the new mirror fitted in, and the individual rim pieces folded back. Only the closest of examinations will reveal that the rim has been cut into several parts.

Occasionally a microscope mirror needs resilvering rather than replacing. The old silver can be removed by strong nitric acid. The quantities of material needed to resilver a microscope mirror are so small that it is perhaps too fiddling a job to do oneself. Nevertheless, silvering is not a task outside the scope of an amateur.

Although the quantities below are small, they will silver four or five microscope mirrors. First of all, the following solutions are prepared: (1) 22 grains of nitrate of silver, 1oz of distilled water; (2) $\frac{1}{4}$oz pure caustic potash, 6oz of distilled water; (3) $\frac{1}{8}$oz milk sugar in power, $1\frac{1}{4}$oz of distilled water. Take $\frac{1}{2}$oz of solution (1), add ammonia drop by drop until the precipitate first formed is just dissolved, add 1oz of solution (2), and then ammonia again until the solution becomes clear. Make up to 4oz with distilled water, then add solution (1) drop by drop until a slight grey precipitate appears. When this has settled, add $\frac{1}{2}$oz of solution (3), and stir well. The mirror should be levelled until it is exactly horizontal, perhaps placed on a builders' spirit level, and then the solution should be carefully poured on to it. This is the first stage. The mirror is left in a moderately warm place free from dust, and after a few hours the liquid should be poured off, and a second lot poured on. The silver should now be thoroughly deposited, and after another few hours the mirror should be lightly rinsed in distilled water, allowed to dry, and perhaps protected by a coat of thin paint, though the mount, of course, will protect the silver.

The price of optical glass is now a fraction of what it was during the great days of instrument making, and this enables

CLEANING AND RENOVATION OF SCIENTIFIC INSTRUMENTS

microscopes to be put on the market for ridiculously small amounts of money. The collector of scientific instruments should not scorn the lenses of modestly priced Japanese microscopes. For the more adventurous, there is another way to obtain lenses to replace those that are missing or broken—spectacle lenses. The 'pebbled' spectacle lenses can be picked up for next to nothing at any junk stall, and after being boiled three or four times their brittleness is reduced and they are easy to grind down. One old London craftsman has been guarding this secret for many years, knowing full well that this is one of those things people think about but never do, a talking point for theorists. That he does do it, and makes first-rate microscope lenses from these discards, may encourage others to follow suit.

When dealing with mirrors or lenses one is inclined to think solely of microscopes, but one must remember that certain of the renovations and repairs mentioned above have an application to the telescope. The silvering of mirrors is also of more than academic interest when it comes to refurbishing surveying instruments that use mirrors, and lenses are used in a wide variety of instruments for reading Vernier scales.

Amongst dealers in scientific instruments there are surprisingly few who are prepared to repair and thoroughly recondition their pieces, and although brass is not a difficult metal to work most dealers fight shy of making any missing parts, even of finding screws to bring, say, a microscope up to working order. This unwillingness to involve oneself in time-consuming repairs is not confined to dealers in instruments; dealers in furniture often prefer to turn a damaged piece over quickly rather than carry out simple repairs, within the scope of anyone who can handle a plane and a chisel.

Dealers in scientific instruments excuse themselves on the grounds that the collectors, whom they admit are frequently more knowledgeable than they, can pick out a repair or a replacement part and would much prefer an instrument that is in an original, rather than restored, state. Obviously discriminating collectors do not want a bodged job, but where a repair is carried out discreetly and in the spirit of the piece then there are surely many who would acquire an item rather than let it lie around for another five years until a replacement part turns up from a damaged article. Most dealers have boxes of bits, including microscope lenses, screws, and mounts, and they rely on these to provide replacement parts. It is often not realised, even by dealers in a big way of business, that many of the smaller components can be found in model engineering kits. Dealers and collectors who are not themselves skilled in metalwork would do well to read the small advertisements columns of the *Model Engineer* (founded 1898), where owners of small workshops invite queries.

GLOSSARY OF SCIENTIFIC INSTRUMENTS

A feature of this glossary of scientific instruments is not that there are so many instruments, but that, basically, there are so few 'classic' instruments. This is true even in surgical instruments, for besides specialised instruments for investigating the sense organs many of the others arise from the need to find out what is happening in internal organs or in passages such as the urethra. The basic tools, forceps, probes, saws, scalpels, have remained unchanged over the centuries.

Many of the instruments in the glossary will be found to be variations of a type. Instruments for testing milk, indigo solutions, water, may be classified under one heading. Spelling varies enormously from one reference book to another, and where there is confusion I have taken as the authorities the *Oxford English Dictionary* and the *Encyclopaedia Britannica*, which are a good deal more reliable both in definitions and in orthography than so-called dictionaries of scientific instruments. For instance dichroscope can be spelled dichroioscope, and diffusiometer diffusionometer, but where there is doubt both versions have been put in, necessary because a change of one letter can mean a different instrument. A diaphanometer measures the transparency of air, but a diaphonometer measures the transparency of spirits.

A good deal of this confusion lies at the hands of over-enthusiastic scientists and inventors who found a Greek or

Latin root and stuck an affix on, adding an -io or an -ono at will. Naturally this glossary cannot hope to encompass all scientific instruments, especially the one-off variety, but because inventors were orientated to Latin and Greek roots, the use of an instrument can be ascertained by reference to a good etymological dictionary.

Where possible dates have been added to give the glossary an added interest, but in some cases this has not been possible. Where a natural root has been found, affixes have been added ad lib. An example is the Greek word *kystis,* low Latin *cystis,* the bladder, which has not only given currency to the cystoscope, a medical instrument for the examination of the bladder, but to cystitis, cystocele, and cystoma. The unreliability of word-makers is evidenced by further outcrops from *kystis.* At some stage someone decided that cyst, a sac containing morbid matter, would be a good word, disregarding the fact that such a sac had no relevance to the bladder.

The dates of various of the instruments can at times be quite startling, for who would have believed that as early as 1825 scientists were measuring the amount of dew on an object with a drosometer, or that fibre particles were being measured with the eriometer (or erinometer) in 1829? It is even more astonishing to realise that the first recorded use of the word periscope occurred as late as 1899, though the word was used in connection with camera lenses in 1865.

A little harmless entertainment may be obtained from wondering why some of these instruments were invented. What surveyor would want a quintant, a rare angle-measuring instrument, when he already had sextants, quadrants, and octants? It is surely commonsensical that instruments measuring a quarter, a sixth, and an eighth of a circle would be sufficient to deal with everything a quintant (a fifth of a circle) could do.

Representative instruments terminating in both -graph and -meter have been put in the glossary, but others have been omitted as all that is involved is the addition of the substitution of a recording device (revolving drum and pencil etc) for a

GLOSSARY OF SCIENTIFIC INSTRUMENTS

measuring gauge or dial. Where there could be confusion, as in aerometer, aerophone, and aeroscope, all of which deal with totally different things (one with air and gas densities, one with sound amplification, and the other with bacteria and spore collection from the air) then variations rising from the same root have been put in.

Absorptionmeter	Measures the absorption power of glass.
Accelerometer	Determines the acceleration of a moving body.
Acetimeter (1875)	Estimates the amount of acetic acid in liquids.
Acetometer	A hydrometer graduated to determine the strength of acetic acid.
Acidimeter (1839)	Determines acidity in cheese or butter.
Ackermann's calculating disc	Determines alcohol and extract in beer.
Acoumeter (1847)	Measures the acuteness of the sense of hearing.
Acousimeter	As above.
Actinograph	Records the variations of the chemical influence of the sun's rays.

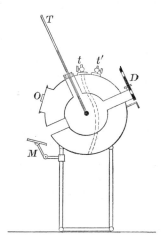

100 An actinometer, for measuring chemical and atmospheric changes.

GLOSSARY OF SCIENTIFIC INSTRUMENTS

Actinometer	Measures the variations of the chemical influence of the sun's rays.
Acupuncturator	Used in acupuncture.
Acutometer	Determines the degree of vision; a device whereby a straight black line is gradually widened.
Adeney's apparatus	Experiments with and analyses dissolved gases.
Adjusting cone	Measures the distance between the axes of the eyes.
Aerometer	Measures the weight and density of air and gases.
Aerophone	A device for amplifying sound.
Aeroscope	Collects spores, bacteria, etc., from the air.

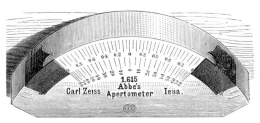

101 An apertometer for measuring the aperture of a lens.

Aesthesiometer (1851)	Determines at how short a distance apart two impressions on the skin can be recognised. In essence, two pricking instruments set at a known distance apart.
Aethrioscope (1832)	Determines the radiation of the sky.
Air meter	Measures the velocity of air currents.
Air poise	Measures the weight of the atmosphere.
Albumenometer	Determines the proportion of albumen in urine.
Alcoholometer (1859)	A hydrometer graduated to indicate strength of alcohol in spirits.
Aleurometer (1844)	Measures the elasticity of the gluten of flour.
Alkalimeter	Measures the strength of alkali in a mixture.

GLOSSARY OF SCIENTIFIC INSTRUMENTS

Altimeter (1847) — Measures altitudes.
Altiscope — Lenses and mirrors arranged to see over intervening objects, a periscope.
Amblyoscope — Tests the muscles of the eye.
Ambulator (1859) — A surveying instrument for measuring linear distances, by means of a geared wheel and a dial.
Ammeter (1882) — A galvanometer for the measurement of electric currents in amperes.
Ammonia gauge — A pressure gauge measuring vapour pressure of ammonia.
Ammoniameter — Measures the strengths of ammonia solutions.
Anemograph — A recording wind-gauge.
Anemometer (1727) — Measures the force of wind, air currents, gases.

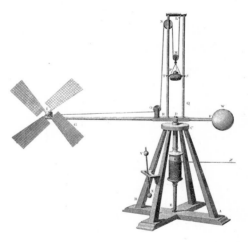

102 An anemometer for measuring wind force.

Anemoscope — Indicates the direction of the wind.
Aneroidograph — A self-recording aneroid barometer.
Anglemeter — Measures the dip of geological strata.
Angleometer — Measures external angles.
Anorthoscope — An optical device for producing images by two revolving discs.
Anthrocometer (1847) — Measures the carbonic acid in a mixture.

GLOSSARY OF SCIENTIFIC INSTRUMENTS

103   Another type of anemometer.

Antimeter (1819) A quadrant for measuring very small angles.
Apertometer (1880) Measures the aperture of a lens.
Aphengoscope A magic lantern for exhibiting opaque objects.
Apomecometer (1869) Measures the heights of objects at a distance.
Apophorometer Enables sublimates obtained from substances at high temperatures to be weighed.
Arcograph (1822) Describes an arc without the use of a central point as in compasses.
Argentometer (1879) Measures the density of silver solutions.

GLOSSARY OF SCIENTIFIC INSTRUMENTS

Arithmometer — An obsolete name for a calculating machine.

Armillary sphere — An astronomical device composed of rings representing the circles of the celestial sphere.

Armillary sphere sundial — A skeleton sphere in which the shadow of the central pole falls on the divided equatorial circle and indicates the time.

Artificial horizon — A device giving the horizon when it is obscured.

Aspirator (1863) — Passes air or gases through or over certain liquids or solids.

Assay apparatus — Analyses metallic ores etc.

Astrodicticum — An instrument that permits many people to view the same star instantaneously.

Astrolabe — The earliest instrument for taking the altitude of a star, rendered obsolete by the sextant.

Astrolage — A nautical astrolabe.

Astrometer (1830) — Invented by Herschel for measuring star magnitudes.

Astrometeoroscope — Demonstrates the effects of persistence of vision.

Astropatrotometer — Measures the brightness of a star by comparison with a norm.

Astroscope (1675) — Two cones on which the constellations were depicted.

Atmidometer (1830) — Measures the evaporation of ice and snow, etc.

Atmolyser (1866) — Illustrates separation of gases.

Audiometer (1879) — Measures the power of hearing against a scale.

Audiphone (1880) — Conveys sound to the auditory nerve by being placed against the teeth.

Auriscope (1853) — Examines the inner ear.

Auxanometer — Measures plant growth, usually with recording device.

Axometer (1865) — An optician's instrument to find the optimum height of the bridge of a pair of spectacles.

Azotometer	Measures the proportion of nitrogen in air or a mixture.
Back staff	A navigational instrument similar to the cross staff.
Bacterioscope (c1886)	Examines bacteria.
Balance, chemical	A very sensitive balance, consisting of a horizontal beam supported on a knife-edge on top of a pillar. Pans are suspended from ends of beam or knife-edges.
Balance, hydrostatic	Weighs substances in order to ascertain specific gravity.
Balance, physical	A less sensitive chemical balance for weighing heavier things.
Balance, specific gravity	A spring balance to find out specific gravities. The substance to be weighed is suspended from a spiral spring.
Balance, spring	Weighs objects by the extension or compression of a spiral spring.
Barograph (1865)	A type of aneroid barometer, with clockwork-driven drum, pen and chart. Atmospheric changes are recorded in the form of curves.
Barometer (1665)	Determines the weight or pressure of the atmosphere by means of mercury.
Barometer, aneroid (1849)	A barometer without liquid. It works by the effect of varying pressures on the elastic sides of a metal box from which the air has been extracted.
Barometer, Boylean Mariotte	A portable mercury barometer.
Barometer, Collie's	A portable mercury barometer in which the usual glass tube is replaced by a flexible rubber tube.
Barometer, dial or wheel	A mercury barometer in which the movement of the mercury is measured by a float and counterpoise fixed to a needle rotating over a dial.
Barometer, Fitzroy	A design of barometer made famous by Admiral Fitzroy.
Barometer, Fortin	A portable barometer with a flexible cistern.

GLOSSARY OF SCIENTIFIC INSTRUMENTS

Barometer, glycerine — Uses glycerine instead of mercury.
Barometer, King's — A self-recording mercury barometer in which the mercury is weighed rather than measured against a scale.
Barometer, marine — A barometer with a constricted tube suspended rather than hung.
Barometer, self-recording — Records changes of pressure on a dial.
Barometer, water — Variations of pressure are measured by the changes in height of a column of water.
Baroscope (1881) — Demonstrates that bodies in air lose as much of their weight as that of the displaced air.
Bathometer (1875) — Measures the depth of the sea.
Batoreometer — Measures minute thicknesses. The contact of a micrometer screw with the object to be measured is shown by the passage of an electric current.
Battery, gravity — Different liquids range themselves at various heights depending on their specific gravities.
Beam compass — A wooden or metal beam with sliding sockets holding steel or pencil points, used for describing large arcs out of range of ordinary compasses or for measuring small distances exactly.
Besidometer — An optician's instrument measuring spectacles from hinge to hinge.
Bevel protractor — A flat protractor for setting off angles.
Binnacle (1662) — The compass housing on board ship.
Binocular (1871) — Twin-tubed telescope, for the convenience of both eyes. In the plural, field glasses.
Blanchimeter (1847) — Measures the bleaching power of chloride of lime and potash.
Bolometer — Measures minute quantities of radiant heat.
Boning rods (c1785) — Short T-squares used in surveying.
Bow compass (1796) — For describing small arcs.
Brook's curves — Flexible steel bands providing curves for draughtsmen.

GLOSSARY OF SCIENTIFIC INSTRUMENTS

Burette (1836)	Graduated glass tube and stop-cock for delivering measured quantities of liquid or gas.
Butyrometer	Determines the proportion of fats in milk.
Cadran	Measures the angles of gems when being polished or cut.
Calcimeter	Analyses substances containing lime.
Calliper square	Draughtsman's square, with a graduated bar and adjustable jaws.
Callipers (1588)	Measures the external or internal diameters of an object.
Calorescope	Produces heat for laboratory purposes.
Calorimeter (1794)	Measures the quantities of heat received by or from bodies, or developed or absorbed by combination, combustion, friction, etc.
Campylometer	A cartographic aid for reading off lengths on maps or plans.
Carbolimeter	Measures carbolic acid.
Carbonometer	Detects the presence of carbonic acid.
Carburometer	Determines the gases contained in carbonic oxide, etc.
Cardiograph	Records the action of the heart.
Cathetometer (1864)	A horizontal telescope moving on a graduated vertical support for measuring height differences.
Catopter (1644)	A mirror, a reflecting optical instrument.
Catoptric dial	Indicates the time by reflection in a mirror.
Catoptric light	A mirror or series of mirrors reflecting light from one source in a parallel beam.
Centrolinead	Draws lines in perspective.
Cephalometer	A medical instrument that measures the head of an unborn child.
Chains, land	Used to establish base lines in surveying.
Chart dividers	Used for taking measurements on charts.
Chartometer	Measures distances on charts.

GLOSSARY OF SCIENTIFIC INSTRUMENTS

Cheilvangroscope — Examines the circulation of the blood.

Chlorometer (1826) — Tests bleaching powers of chloride of lime.

Chorograph (1839) — Constructs triangles in surveying.

Chromascope — Demonstrates optical effects of combining colours.

Chromatometer — Measures the degrees of colour.

Chromatoscope — An eccentrically rotating reflector telescope, used in observing the scintillations of a star.

Chromatrope (1860) — By using coloured rotating discs, phenomena dependent on the persistence of vision were demonstrated.

Chromometer — Determines the colour of petrol, or the presence of minerals in ores.

Chronograph (1868) — Used for recording time by means of a puncture or mark on a moving paper strip.

Chronometer (1735) — An accurate timepiece for use in navigation and astronomy.

Chronopher (1867) — Signals the time to distant places by electricity.

Chronoscope (1794) — Measures minute intervals of time.

Chronothermometer — A circular thermometer with a watch movement in the centre, with a hand that revolves over the thermometer at a rate of 20°F in 15 minutes. Used for testing petrol by measuring the heat of a flame.

Chyometer (1880) — A tube containing a graduated piston, used for measuring liquids.

Circle (1752) — Surveying and astronomical instrument.

Circumferentor (1610) — A surveying instrument for measuring vertical and horizontal angles, in the form of a circle. Rendered obsolete by the theodolite.

Clepsydra (1646) — A type of water-clock, measuring time by the transfer of water from one vessel to another.

Climatometer — Measures fluctuations of temperature in the body.

Clinograph An adjustable set-square used in surveying.

Clinometer (1811) Used for measuring angles of inclination by reference to a plumbline or a level.

104 Clinometer, used for measuring angles.

Clinometer, Abney's Measures angles of inclination by means of reflection.

Clinometer compass A clinometer sight combined with a compass.

Cloud mirror A black glass mirror marked with the compass points and aligned on adjustable screws for charting the positions of clouds.

Collimator (1865) A telescope arranged to transmit a parellel beam of light, and part of the spectroscope.

Colorimeter (1863) Measures depth of colour, especially in liquids.

Comparator (1883) Accurately measures lengths or checks graduated scales.

Compass Indicates directions upon the earth's surface by means of a magnetic needle.

Compass, azimuth A compass divided into degrees with vertical sights.

Compass, liquid A compass with the card floating in liquid.

GLOSSARY OF SCIENTIFIC INSTRUMENTS

105 Colorimeter, used for measuring depth of colour.

106 Azimuth compass, a compass provided with vertical sights.

Compass, marching	A military compass with a movable indicator that can be set to the angle between the meridian and the marching direction.
Compass, overhead	A compass read from beneath.
Compass, Paget	A type of compass with short needles.
Compass, prismatic	A compass with a prism for viewing the card and a distant object at the same time.

GLOSSARY OF SCIENTIFIC INSTRUMENTS

Compass, projector   A ship's compass which projects an image on a screen.
Compass, spirit   A liquid compass in a liquid other than water.
Compass, vertical card   A compass with a prism arranged to produce a vertical image.
Compasses   Two limbs rotating on the same axis, one ending in a point, the other in a pencil.
Compasses, proportional   Compasses used in reducing or enlarging designs, with two pairs of points.
Compasses, triangular   Compasses with three legs.
Compensator (after 1837)   A device that neutralises the effect of iron on a ship's compass.
Compressor   A microscope fitting that holds live objects.
Computing scale   A rule with a sliding frame for estimating map areas.
Conchometer (1828)   Measures sea-shells.
Condenser (1798)   A lens system for concentrating light at one point.
Cosmolabe   A form of astrolabe.
Cosmosphere   A hollow glass globe with model of the earth inside and constellations depicted outside.
Cross head (1874)   A brass octagon with sights on each face used by surveyors for setting out angles.
Cross staff (1669)   Shaped like a cross and used to measure the angle of the altitude of the sun above the horizon. A basic early navigational instrument.
Cryophorus (1826)   Illustrates the freezing of water by evaporation.
Curve tracer   Records curves in conjunction with a galvanometer.
Cyanometer (1829)   Measures degrees of blueness in the sky.
Cycloidograph   Draws cycloids.
Cyclometer (1815)   Measures circular areas.
Cymograph (1837)   Takes profile drawings from a model in bronze or brass.

GLOSSARY OF SCIENTIFIC INSTRUMENTS

Cystoscope — A medical instrument for the examination of the bladder.

Dasymeter (1872) — Tests the density of gases, consisting of a thin glass globe.

Declinometer (1883) — Measures the magnetic declination of any place when the meridian is known.

Deflector (1837) — Compares the magnetic directive force on different courses on board ship. Usually an adjustable magnet fitted for application to the compass bowl and rotating about the compass centre.

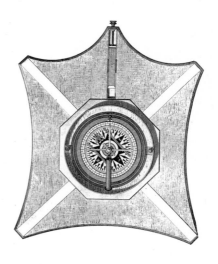

107 Compass fitted with corrector to counter effects of magnetism.

Delineator (c1782) — A surveyor's instrument on a geared wheel. An ambulator or waywiser.

Demi-circle — Another name for the protractor.

Dendrometer — Determines the height and diameter of trees.

Densimeter (1863) — Ascertains specific gravity of a liquid.

Diagonal scale — A rule depicting parallel lines with other lines crossing them at equal angles and at equal distances from each other.

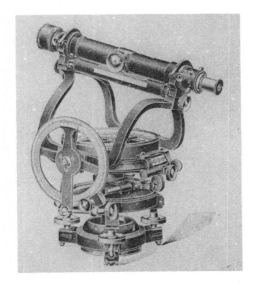

108 Mining dial, a form of underground theodolite.

Dial, mining (1523)	A form of magnetic compass, with sights, used in mining, and underground surveying.
Diaphanometer	Measures the transparency of the air.
Diaphonometer	Measures the transparency of spirits.
Dichroscope (1857)	Exhibits the complementary colours of polarised light, and used in the examination of precious stones.
Diffusiometer (1866)	Measures the rate at which diffusion of gases takes place.
Dilatometer (1882)	Measures the expansion of a substance, especially a fluid.
Dipleidoscope (1843)	Determines the apparent time of noon, consisting of two mirrors and a plane glass arranged as a prism. By the reflection of the sun's rays two images reach the eye, merging into one when the sun's centre is on the meridian.
Director	A military theodolite.

GLOSSARY OF SCIENTIFIC INSTRUMENTS

Distance finder — Used for calculating off-shore distances at sea and also for keeping ships on station, as in a convoy. Consists of a moveable prism plate, a diagram plate and a sighting telescope.

Dividers (1703) — A form of compass with two metal points.

Dividing engine — A machine for laying out a circular and linear matrix for scientific instruments.

Dosimeter (1881) — Measures the number of drops.

Drosometer (1825) — Measures the amount of dew on an object.

Ductilemeter — Measures the ductility of metals.

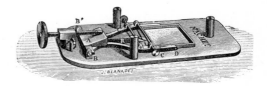

109 A compressor, a microscope fitting that holds live objects etc.

Dynactinometer (1851) — Measures the intensity of light-producing rays and determines the power of lenses.

Dynameter (1828) — Measures the magnifying power of telescopes by applying a micrometer to the eyepiece and measuring the diameter of the pencil of light.

Dynanometer — Measures muscular strength.

Ebullioscope (1880) — Ascertains the boiling point of liquids.

Eccentrolinead — An instrument for drawing eccentric lines.

Echometer (1736) — Measures the duration of sounds.

Echoscope — A medical instrument that amplifies sounds produced by percussion of the thorax.

Eclipsareon — An obsolete instrument for demonstrating eclipses.

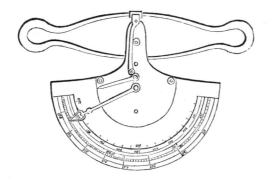

110 A dynanometer, constructed on the principle of a fairground try-your-strength machine.

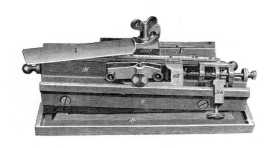

111 A microtome, for freezing and cutting specimens for microscopes.

Eidograph (1801)	A form of pantograph.
Eidoscope	A form of kaleidoscope, using two perforated metal discs revolving on independent axes.
Eikonometer	Determines the magnifying power of a microscope.
Elacometer	Decides the proportion of oil in a given substance.

GLOSSARY OF SCIENTIFIC INSTRUMENTS

Electrometer (1749) — Measures differences of electrical potential.

Ellipsograph — Used for drawing ellipses.

Endoscope — A medical instrument for examining the rectum.

Engiscope (1684) — A type of microscope made on the reflector pattern, with the eyepiece in the side of the tube.

Eriometer (1829) — Measures the diameter of fibre particles.

Esthesiometer (1851) — Measures tactile sensibility in the human body.

Evaporimeter (1828) — Measures the quantity of liquid evaporated in a set time.

Faciometer — Measures the face for spectacle-fitting.

Field glass (1831) — A binocular telescope in handy form.

Fluxmeter — Used in conjunction with a galvanometer in the exploration of magnetic fields.

Focometer (1853) — Measures the focal length of a lens.

Fore staff (1669) — A navigational instrument allied to the cross staff.

Galactometer — Determines the specific gravity of milk.

Galvanometer (1802) — Measures electric currents by the deflection of a magnetic needle in the magnetic field created by an electric current, or by the deflection of a moving coil. A variation is the mirror galvanometer in which the current deflects a mirror.

Galvanoscope (1832) — Detects the presence of feeble electric currents.

Galvanothermometer — Measures the heat of a galvanic current.

Geodescope — A combined celestial and terrestial globe.

Geometric square — A portable surveying instrument for finding angles, in the form of a square frame. Much less used than circular or semi-circular surveying instruments.

GLOSSARY OF SCIENTIFIC INSTRUMENTS

Geothermometer	Measures the temperature of the earth at various depths.
Globulimeter	Measures the number of red corpuscles in the blood.
Glucometer	Measures the specific gravity and quantity of sugar in grape or other fruit juice before fermentation.
Goniometer (1766)	Measures the angles of crystals.

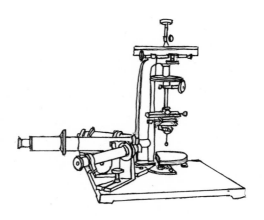

112 A goniometer for measuring the angles of crystals.

Gradimeter	A form of surveyor's level.
Gradiometer	As above.
Gradioplane	A surveying instrument for underground work.
Graduator	A device for dividing lines, straight or curved, into equal parts.
Granometer	An instrument for counting seeds.
Graphometer (1696)	A semi-circular surveying instrument for angle measuring.
Gravimeter (1797)	Measures the specific gravity of substances.
Groma	An ancient device for establishing right angles.

GLOSSARY OF SCIENTIFIC INSTRUMENTS

Gunter's scale (1706)	A wooden scale 2ft long used in surveying and navigation. On the reverse are logarithm tables.
Gyroscope (1856)	A rotating wheel mounted in a ring or rings demonstrating the dynamics of rotating objects.
Gyrostat (1879)	A more sophisticated gyroscope.
Haedromograph	Registers the velocity of the blood.
Haemachrometer	Measures the proportion of haemaglobin in a fluid.
Haematometer (1854)	Counts the number of corpuscles in a set quantity of blood.
Haemoscope	Measures the thickness of a layer of blood for spectroscopic observation.
Halometer (1854)	A goniometer, measuring angles of crystals.
Haloscope	A demonstration instrument exhibiting halos, etc.
Hanging dial	A type of compass used underground.
Harmonometer (1823)	Measures the harmonic relationships of musical notes.
Helicograph (1851)	A drawing instrument for drawing spiral lines on a plane.
Heliochronometer (1875)	A form of sundial.
Heliometer (1753)	An astronomical instrument that measures distances and relative directions of two stars.
Helioscope (1675)	A telescope fitted with smoked lenses for solar work.
Heliotrope (1822)	An astronomical device that shows when the sun has arrived at its furthest point north or south.
Holometer (1696)	Measures angles.
Holophote (1859)	A lamp with lenses or reflectors that collect light rays and throw them in a set direction.
Horometer (1775)	An instrument for measuring the time.
Hyalograph	An instrument for tracing a design on a transparent surface.
Hydrodynamometer (1890)	Measures the velocity of water.

GLOSSARY OF SCIENTIFIC INSTRUMENTS

Hydrometer (1675)	Determines the specific gravity of liquids. Made in many varieties such as the alcoholometer hydrometer, barktrometer hydrometer, freeboard hydrometer, etc.
Hydrophore (1842)	A device for obtaining water from any desired depth for analysis.
Hydropyrometer	A form of pyrometer in which metals, etc, are exposed to great heat then quenched in water, the temperature of which is then measured.
Hydroscope	Detects the presence of water in air; also a type of water clock; also an underwater telescope.
Hygrograph (1670)	Records changes of humidity in the atmosphere.
Hygrometer (1670)	Measures changes of humidity.
Hygrophant	A hygrometer giving instant results, without resort to tables.
Hypsometer (1840)	A device for determining altitudes by reference to the boiling point of water.
Hysterisimeter	Measures strain in materials.
Inclinometer (1842)	Measures vertical magnetic force.
Indigometer (1828)	Measures the strength of indigo solutions.
Indiscope	A type of ophthalmoscope.
Inductometer (1839)	Measures the force of electrical induction.
Interferometer (1899)	Measures minute distances in terms of the wavelength of light, consisting of four plates of glass, or two plates, with portions silvered.
Isograph (1872)	An adjustment set-square.
Kaleidoscope (1817)	An instrument containing fragments of coloured glass etc., and mirrors, so arranged that the disordered fragments appear to form into symmetrical patterns.
Keratometer	Measures the curvature of the cornea of the eye.
Keratome	A surgical instrument for dividing the cornea of the eye.

113 A keratometer, which measures the curvature of the cornea of the eye.

Klinostat	An obscure instrument for determining the influence of gravitation on plant growth.
Koniscope	A device for testing the amount of dust in the air, comprising an air-pump and a tube.
Kymoscope (1867)	Demonstrates sound-wave motion in acoustics.
Lactobutyrometer	Measures the amount of fats in milk.
Lactometer	A galactometer, for measuring specific gravity of milk.
Laryngoscope (1860)	A surgical instrument comprising two mirrors and a light for examining the larynx.
Lencoscoep	Tests the colour perception of the eye.
Level	With the theodolite the most important surveying instrument, used to find a horizontal.
Level, Abney	A variation, comprising a spirit level, a divided arc, and a sighting tube. A reflector in the sighting tube enables the level bubble to be seen at the same time as the object.
Level, Aldis optical	A level in which the bubble is magnified and used as a lens.
Level, Dumpy (c1848)	A level in which the telescope is fixed.
Level, reflecting	A level incorporating a small mirror.
Levelling staff	A post or pole, often telescopic, marked with divisions, used in conjunction with theodolites and other surveying instruments.

GLOSSARY OF SCIENTIFIC INSTRUMENTS

Linen prover — A simple magnifying instrument for counting the threads in a given piece of linen.

Lithotrite (1839) — Crushes stones in the bladder.

Litrameter — Measures the specific gravity of liquids.

Logograph (1879) — Represents on a moving paper strip the motion of air waves of speech, and consists of a horn and a diaphragm, plus the recording mechanism.

Logometer (1813) — A scale for measuring chemical equivalents.

Lucimeter (1825) — Measures the intensity of light; a photometer.

Lysimeter — Measures the water that percolates through a given depth of soil.

Macrometer (1825) — Measures the size of distant objects by means of two reflectors affixed to a sextant.

Magnetograph (1847) — Records the states and variations of terrestrial magnetic phenomena.

114 A magnetometer, used for measuring magnetic declination sometimes in association with a declinometer.

GLOSSARY OF SCIENTIFIC INSTRUMENTS

Magnetometer (1827) — Measures magnetic declination; essentially a needle swinging in a horizontal plane.

Magnifier (1550) — A lens or lens system set in a frame, usually with a handle or on a stand. The triple anaplat magnifier has two lenses of flint glass separated by a double convex lens of crown glass.

Manometer (1730) — Measures the tension or force of gases etc., by allowing the gas to exert pressure against a tube of mercury, against air or other gas in a closed tube, or in bending a tube or spring to activate an index.

Meanometer — A wooden scale with a sliding index.

Meatoscope — A medical instrument for examining the urethra. A type of speculum.

Megalethoscope — A stereoscopic viewer.

Megameter — Determines longitude by astral observation.

Megascope (1831) — A form of magic lantern.

Mekometer (1894) — A military range-finder comprising two angle-measuring instruments connected by wire and operated by two men.

Melanoscope — An optical device containing coloured glasses used to view flames to detect potassium, etc.

Meridian circle — An astronomical circle fastened at right angles to a horizontal axis and turning with it.

Meteorograph (1780) — A combination instrument composed of a barograph, thermograph, hydrograph, and wind-velocity recorder formerly sent up into the upper atmosphere by means of ballons or kites.

Meteorometer — An apparatus installed to receive temperatures, atmospheric pressures, rainfall etc.

Meteoroscope — Used for taking angles and measuring heavenly bodies.

Metrochrome — Measures colours.

GLOSSARY OF SCIENTIFIC INSTRUMENTS

Metroscope (1855) — An instrument for examining the womb.

Micrograph — A pantograph for graphically repeating small writing etc.

Micromanometer — A form of manometer.

Micrometer (1790) — An instrument used with a telescope or microscope for measuring small distances, or a tool for measuring small lengths in engineering or mechanics.

Micrometer, bifilar — A micrometer with two threads, one of which at least is moveable.

Micrometer, calliper (1884) — A calliper with a micrometer screw.

Micrometer, cobweb — An optical micrometer using cobweb threads as a guide in the manner of a gun-sight.

Micrometer, dioptric — A double-image micrometer. The two halves of a bisected lens are moved along their line until the image blends. Distances are measured by the number of screw turns needed to bring this about.

Micrometer, double image — As above.

Micrometer, double refraction — A type of double-image micrometer.

Micrometer, filar — A cobweb micrometer.

Micronometer — Measures very short times.

Microscope (1656) — An optical instrument consisting of a lens or lens system mounted in a tube for viewing small objects.

Microscope, binocular — A microscope with two tubes, and one objective, giving binocular vision.

Microscope, reflector — A microscope incorporating mirrors instead of an objective.

Microscope, solar — A microscope using sunlight to illuminate the specimen.

Microspectroscope (1867) — Combination of microscope and spectroscope.

Microtome (1856) — A microscope accessory for freezing and cutting specimens.

GLOSSARY OF SCIENTIFIC INSTRUMENTS

Mirrorscope	A form of magic lantern for viewing opaque subjects.
Moderator (1851)	A microscope accessory, used to diffuse light passing from a lamp to the stage.
Monochromator	An obsolete name for the spectroscope.
Mydynamometer	Measures the muscular strength of a man.
Myograph (1867)	Takes tracings of muscular contractions and relaxations.
Myrioscope	A type of kaleidoscope.
Napier's bones (1658)	A set of rods used to assist calculations, a kind of pocket abacus.
Natrometer	Measures the percentage of soda in salts of potash and soda.
Nauropometer	Measures the amount a ship heels at sea.
Nautigan	A navigational calculating aid.
Navigraph	A form of sextant.
Navisphere	A navigational celestial globe.
Nephoscope (1881)	Measures the distances or velocities of clouds.
Night glass (1779)	A telescope with large object glass.
Nitrometer (1828)	Measures the amount of nitrogen in a compound.
Nocturnal (1627)	An early navigation instrument made to find the latitude, aided by fixed stars, and thus the time.
Object staff	A levelling staff.
Octant (1661)	Measures angles; the eighth part of a complete circle.
Odograph	A pedometer.
Odometer	As odograph.
Odontograph (1838)	Used for marking out the teeth of gear wheels in engineering.
Oenometer	A variety of hydrometer used for finding the alcohol content of wines.
Oleometer	A variety of hydrometer for finding the densities of oils.
Ombrometer	A rain gauge.
Omnimeter	A theodolite with an extra scale for measuring distance.

GLOSSARY OF SCIENTIFIC INSTRUMENTS

Oncometer — A medical instrument for studying variations in the size of the kidney, the liver, etc.

Opeidoscope (1873) — An instrument comprising a tube linked with a flexible membrane, and a small mirror for showing on a screen vibrations caused by sound.

Operameter (1829) — A machine attachment indicating the number of rotations of a shaft or rod.

Ophthalmoscope (1857) — Used for viewing the interior of the eye.

Opisometer (1872) — Measures a curved line on a map.

Opsiometer (1842) — An optician's tool for measuring distinct vision.

Optometer — See opsiometer.

Orientator (1875) — Used in setting out the ground plan of a church so that the chancel is orientated to the east.

Orograph (1846) — Maps out undulating or hilly contours.

Orrery (1713) — Illustrates the solar system by the revolution of spheres, operated by wires and gearing.

Orthoscope (1875) — An optician's instrument for examining the outer eye.

Oscillating table — A device for demonstrating the principle of inertia.

Oscillograph (1904) — Records wave form of current and voltage in electric current.

Oscillometer — Measures the angle through which a ship rolls.

Osmometer — Measures osmosis in liquids.

Osteophone — Transmits vibration through the bones of the head to be appreciated as sound by the deaf.

Otacoustic (1643) — An ear-trumpet.

Otheoscope (1877) — Exhibits the repellant action of light or heat in an exhausted vessel.

Otoscope (1853) — An instrument with which to view the inner ear.

Ozonometer (1862) — Measures the percentage of ozone in the air.

GLOSSARY OF SCIENTIFIC INSTRUMENTS

Pachometer — Measures thickness of glass, paper etc; also called a pachymeter.

Pagoscope — A form of thermometer to forecast frost.

Palinurus — An instrument on a gimbal mount, used on board ship, to find the true course without resort to calculations.

Pantograph — Used for copying maps and drawings on the same scale, or larger or smaller.

Panyochrometer — A combination of sundial and compass.

Pantometer — A surveying instrument for measuring angles.

Parabola — A microscope accessory for reflecting light on an opaque object from below.

Paralleloscope — A military device for using gun-sights in a confined space.

Pedometer (1723) — A pace-measurer usually carried in a pocket or hooked to a belt.

Pelorus or Dumb Plate — A compass card or dial without needles on a gimbal mount and fitted with sights.

Pelvimeter — Measures the diameter of the pelvis.

Pendulum (1660) — Used for a variety of purposes, such as for the determination of gravitational force or investigations of motion. There are two basic kinds, the simple pendulum, using string or cotton, and the compound pendulum, incorporating a rigid rod.

Perambulator (1828) — Measures distances by means of a geared wheel and dial.

Perimeter (1875) — An optician's instrument for investigating sensitivity of the retina.

Periscope (1899) — A form of bent telescope using prisms for viewing obscured objects.

Perspectograph — A draughtsman's tool for drawing objects in perspective.

Phanarogrisonmeter — Discerns dangerous gases in mines etc.

Phantoscope (1894) — A kaleidoscope.

Phenakistoscope (1834) — A device for exploring the phenomenon of persistence of vision.

GLOSSARY OF SCIENTIFIC INSTRUMENTS

Phoneidoscope (1878) — Investigates the colour figures of liquid films under vibration.

Phonometer — Measures the number of vibrations of a sound in a set time.

Phonorganon — A speaking machine, using diaphragms, valves etc.

Phonoscope — Used for observing the properties of sounding bodies such as violins.

Phonoshote — Transforms sound into light.

Phosphoroscope (1860) — Investigates the duration of phosphorescence in various substances.

Photodrome — A large wheel with spokes illuminated by light passing through slits in a revolving disc.

Photoheliograph (1861) — A telescope modified to take photographs of the sun.

Photometer (1760) — Measures intensity of light.

Photophone (1880) — Communicates sounds by a beam of light.

Phrenograph — Records the movements of diaphragm, etc, in breathing.

Picket pole — A form of levelling rod used in surveying.

Piezometer (1820) — Measures pressure, especially water, usually a gauge attached to a water supply.

Pioscope — Tests milk.

Plagiograph — A form of pantograph.

Plane table (1607) — Drawing board and alidade for field surveying.

Planetarium (1860) — Demonstrates the movement of the planets.

Planigraph — A pantograph.

Planimeter (1858) — Measures the area of irregular plane figures.

Planisphere — An aid to the study of heavenly bodies.

Planometer — Measures and tests a plane surface.

Platometer — Measures areas on maps and plans.

Platymeter — Measures the inductive capacity of insulated materials.

Plethysmograph (1872) — Registers the difference in size of human limbs.

GLOSSARY OF SCIENTIFIC INSTRUMENTS

Pluviograph — A recording rain gauge.
Pluviometer (1791) — Measures the amount of rain.
Pneometer — A spirometer, measures capacity of human lungs.
Polarimeter (1864) — Measures the rotation of the plane of polarisation in polarised light.
Polariscope (1842) — An instrument consisting of a crystal producing polarised light and an analyser.
Polemoscope — Field glasses with an oblique mirror placed for observing objects that do not lie before the eyes.
Polygonoscope — A kaleidoscope.
Position finder — Used for finding a position on a map or the position of a ship on a chart.
Potentiometer (1881) — Measures the potential between two points in an electrical circuit.
Potometer — Measures the watery vapour exuded from the leaves and stems of plants.
Praxinoscope (1882) — A device or toy using the phenomenon of persistence of vision.
Protractor (1658) — An instrument for laying down and measuring angles on paper, sometimes equipped with a radial arm and a Vernier.
Protractor, chart — A navigational protractor divided in degrees for laying off the ship's course on a chart.
Protractor, isometrical — Constructed for isometrical perspective drawings.
Pseudophone — A laboratory instrument used in investigating acoustics.
Pseudoscope — Exhibits objects with their relief reversed.
Psychrometer (1727) — A type of hydrometer.
Pupilometer — Measures the size of the pupil of the eye.
Pycnometer — Measures the density of liquids.
Pyrheliometer (1855) — Measures the heat from the sun.
Pyrometer (1749) — Measures temperatures higher than those obtainable from a glass thermometer. The simplest form relies on information from the expansion of a

	metal rod or gas. The principal type of pyrometer relies on the electrical current generated by a junction of two different metals or alloys at heat. For temperatures higher than 1400°C radiation pyrometers, using heat focussed by a mirror, are employed.
Quadrant	Measures altitudes, and is used in gunnery, astronomy and surveying. Hadley's quadrant is a variant incorporating mirrors, so that an observer can see two objects at once. A nautical quadrant (a quarter of a circle) is often called a sextant (strictly speaking, a sixth of a circle).
Quintant	A rare angle-measuring instrument akin to the quadrant.
Radiometer (1875)	Consists of four vanes, each blackened one side and silvered the other revolving vertically in a vessel nearly exhausted of air.
Radiophone (1881)	Produces sound by the action of heat or light.
Ramsden chain	A steel chain used in surveying.
Rangefinder	Finds the distance of an object from the observer.
Reflecting circle	Measures altitudes and angles. A circle fitted with a telescope.
Refractometer (1876)	Measures the refraction of light, and determines the refractive index of transparent objects.
Regulator (1758)	A clock establishing a standard.
Repeating circle (1815)	A surveying instrument used as an alternative to the theodolite.
Respirometer	Indicates the watery vapour exuded from plants, a potometer.
Retinoscope (1884)	A type of ophthalmoscope.
Rheometer	Measures the force of an electric current.
Rheoscope	Detects an electric current.
Rheotome	Periodically interrupts an electric current.
Rhinoscope	Examines the nostrils.

GLOSSARY OF SCIENTIFIC INSTRUMENTS

Rhumboscope — A form of station pointer for locating positions on charts.

Rhysimeter (1871) — Measures the velocity of liquids or speeds of ships.

Rhythmometer (1812) — Any instrument for marking time to music, a metronome.

Rotameter — Measures distances on maps by a small wheel connected to a dial.

Saccharometer (1784) — A type of polariscope for testing sugar solutions by polarised light.

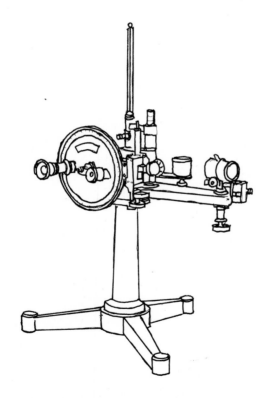

115 A refractometer, measuring refraction, especially refraction indexes of transparent objects.

GLOSSARY OF SCIENTIFIC INSTRUMENTS

Salinometer (1844) — Measures the salt content of water. Also salimeter.

Sceptre recorder — Records the depth of water beneath a ship.

Sclerometer — Measures the hardness of a substance.

Scotoscope (1664) — Displays objects in the dark or in minimum light.

Sector (1598) — Two flat rules hinged together and inscribed with tables.

Seismograph (1858) — Measures and records earthquakes and earth tremors.

Seismophone — Distinguishes underground sounds such as earth movements, underground streams etc.

Selenotrope (1883) — A device showing the phases of the moon.

Semicircumferentor (1712) — A graphometer.

116 The seismograph, used to detect earth tremors. This example was made by James White of Glasgow.

GLOSSARY OF SCIENTIFIC INSTRUMENTS

Set-square	A draughtsman's instrument used in conjunction with a T-square for drawing vertical lines or various angles.
Sextant (1628)	Measures angular distances between objects.
Sextant, box	A portable sextant in a box.
Sextant, double	A sextant able to measure two angles at once.
Sextant, stellar	An astronomical sextant with telescope and mirror.
Sidereal clock	Shows astronomical time.
Sideroscope	Detects minute traces of magnetism by a combination of magnetised needles.
Siderostat (1877)	An astronomical device that keeps a star within the same portion of a telescopic field.
Slide-rule (1663)	Consists of two graduated rules that slide one on top of the other for the instant solution of mathematical problems. More complicated slide-rules have four elements.
Sliding gauge	Used by instrument makers for setting off distances.
Sondograph	Records the configuration of the sea bed.
Sonometer (1808)	Consists of wires or strings stretched over a sounding board. Used to find out the number of vibrations made by a string emitting a musical sound.
Spectrograph	A spectroscope with camera attachment.
Spectroheliograph	A spectrograph used for solar work.
Spectrometer (1874)	A spectroscope fitted with a divided circle.
Spectrophometer	Compares the colours of the spectrum.
Spectroscope (1861)	Consists of a collimator, prism, and telescope. Used for examining spectra.
Speculum (1704)	The mirror used in reflecting telescopes and microscopes, but also a surgical instrument for investigating anal and other passages.

GLOSSARY OF SCIENTIFIC INSTRUMENTS

Spherograph — An instrument used in navigation to locate the position of the ship at any specific time.

Spherometer (1827) — Measures the diameter of spherical bodies and their curvature.

Spheroscope — A model for illustrating astronomy.

Sphygmometer — A medical instrument that measures the pulse rate.

Spinthariscope (1903) — Makes radium emanations visible.

Spirograph — Records the movement of the lungs.

Spirometer (1846) — Measures the capacity of the lungs.

Spring bows — Compasses made from sprung metal, thus dispensing with a joint.

Stadimeter — A horizontal bar on a levelling staff.

Stadiometer (1862) — A surveying instrument used in conjunction with maps.

Station pointer — A surveying instrument for locating positions on charts.

Station staff — A surveying instrument for taking angles in the fields.

Stauroscope (1875) — Used for observing the effects of polarised light on crystals.

Stereometer (1801) — Measures the specific gravity of porous bodies.

Stereomonoscope — An instrument with two lenses that creates a three-dimensional effect from one picture.

Stereoscope (1838) — Gives two pictures a three-dimensional effect by using binocular lenses and two pictures.

Stethometer (1850) — Measures the movement of the chest during breathing.

Stethoscope (1820) — Tests the pulsation of heart and lungs.

Stomatoscope — A medical device for examination of the mouth.

Strabismometer — An optician's instrument for measuring a squint.

Stroboscope (1836) — An optical toy, in which an object appears to move due to light which is periodically interrupted. Also an instrument (1896) for observing the successive phases of a periodic motion

GLOSSARY OF SCIENTIFIC INSTRUMENTS

Sundial

Sunshine recorder

Sympiesometer (1817)

T-square

Tacheometer (1876)

Tachometer (1810)

by means of light periodically interrupted.
An outdoor dial with a gnomon, the shadow of which denotes the hour.
A meteorological instrument recording hours of sunshine.
Measures the weight of atmosphere by compressed gas. A form of barometer.
A drawing instrument for making horizontal lines, used in conjunction with a set-square.
A transit theodolite with a distance-reading gadget in the telescope.
Measures the speed of rotation of a machine.

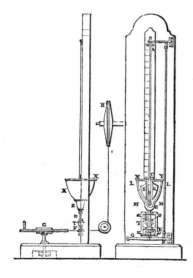

117 The tachometer, a velocity-measuring instrument.

Taseometer

Tasimeter (1878)

Teinoscope (1822)

Measures strains in structures and parts of structures.
Measures very small variations of temperature, humidity, and pressure, etc.
A combination of prisms which increases or diminishes the apparent linear dimensions of objects.

GLOSSARY OF SCIENTIFIC INSTRUMENTS

Teleiconograph   A combination of telescope and camera lucida. The camera lucida by means of prisms or mirrors reflects an image of an object on a flat surface so that it can be sketched.

Telemeter (1860)   A device for estimating the distance of an object from the observer. A range finder.

118 The telemeter, a type of range finder.

Telengiscope   A combination microscope and telescope.

Telepolariscope   A combination polariscope and telescope.

Telescope (1648)   Brings objects nearer by a combination of lenses (refracting telescopes) or mirrors (reflecting telescopes).

Telespectroscope   Another name for an astronomical spectroscope.

Telethermometer   Measures the temperature of a distant object, a form of pyrometer.

Tetanometer   A medical instrument for stimulating nerves by electricity.

Thaumatrope (1827)   An optical device depending for its impact on the persistance of vision,

GLOSSARY OF SCIENTIFIC INSTRUMENTS

	consisting of a card or disc with two different figures on the two sides, when card rotated figures merge into one.
Theodolite (1571)	A surveying instrument for measuring horizontal and vertical angles. A transit theodolite has a telescope that can describe a complete revolution on a horizontal axis. Common theodolites differ little from the circumferentor, and have no telescope, only sights.
Thermoammeter	Measures very weak electrical currents.
Thermograph (1881)	A species of recording thermometer.
Thermometer (1633)	Measures temperature. Most instruments use mercury, but others depend upon changes of electrical resistance. The Beckmann thermometer is accurate to a thousandth of a degree Centigrade, the chemical thermometer has the scale enclosed in glass so that it can be used in caustic substances, the blind-scale thermometer has no scale but only a single mark to indicate a specific temperature.
Thermometograph (1837)	A recording thermometer, especially for the high and low points.
Thermomultiplier	Measures small variations of temperature due to radiant heat.
Thermoscope	Indicates changes of temperature.
Thermostat (1831)	Maintains a constant programmed temperature.
Tintometer (1889)	Used by dyers etc. for optical colour-testing.
Tithonometer	Measures the chemical effects of light.
Tonometer (1725)	An instrument for measuring the rate of vibrations in a sound, or one for investigating the effect of drugs and fluids on organs moved from a dead body, or (1876) an optician's instru-

235

GLOSSARY OF SCIENTIFIC INSTRUMENTS

	ment for measuring tension of the eyeball.
Tonsillotome (1881)	Takes out tonsils.
Topophone	Decides where a sound is coming from.
Trajectograph	Records trajectories in gunnery, etc.
Transit instrument (1793)	A telescope mounted at right angles to a horizontal axis, on which it revolves with its line of collimation in the plane of the meridian; used in conjuction with a clock for observing the transit of a planet etc. over the meridian of a specific place.
Trechometer	A form of odometer. Also known as trocheameter, and trochometer. Measures distances.
Tromometer (1878)	Measures earth tremors.
Turbidimeter	Measures the turbidity of water.
Udometer (1825)	A rain gauge.
Urethroscope	Examines internal passages of the body.
Vaporimeter (1878)	Measures the volume or tension of a vapour. A type of vaporimeter measures the vapour tension of oils.
Velocimeter (1842)	Measures velocity of machines.
Viameter	An odometer.
Vibrometer	Used for measuring the effect of vibrations on delicate instruments.
Vibroscope	Records the vibrations of a tuning-fork.
Vinometer	Measures the strength of wines.
Viscosimeter (1868)	Measures the viscosity of liquids.
Visiometer	Measures the power of spectacles.
Vivascope	A close-range telescope.
Voltameter (1836)	Measures the strength of an electric current.
Voluminometer	Measures the volume of a solid by analysing the difference in tension in or out of confined air.
Waywiser	Measures distances. A geared wheel pushed along the ground.
Zenometer	Measures the angle of depression from a ship's bridge to a buoy or boat,

GLOSSARY OF SCIENTIFIC INSTRUMENTS

Zoetrope (1869)

Zoopraxinoscope

Zymosismeter

thus obtaining the distance. Used in conjunction with a sextant.
An optical device known as the 'wheel of life'. A Victorian toy.
An optical device depending for its effect on persistence of vision, using a revolving cylinder. The same as a zoogyroscope.
Measures the degree of fermentation arrived at by mixing different liquids.

NOTES ON PRICES

PRE-EIGHTEENTH-CENTURY scientific instruments are usually outside the price range of the average collector. Gilt-metal quadrants and other navigational equipment of early times fetch more than four figures, and although diptych dials made in Augsberg have generally made between £600–£800, in Spring 1973 a dial by Martin of Augsberg was sold at Sotheby's for £2,400. Flat dials of the Butterfield type are usually cheaper, but apparent bargains should be closely examined in view of the many forgeries that are being made, especially in France.

119 A Butterfield dial of a desirable type, sold for £250.

NOTES ON PRICES

120 Even in a collectable field such as dials there are objects within the scope of the small collector. This English nineteenth-century diptych dial made only £25 at auction.

Although 1973 has seen a rise in the prices obtained at auction of scientific instruments, eighteenth and nineteenth century navigational and surveying instruments are if anything underpriced. Eighteenth century simple sextants have passed the hundred pound figure, but later, more sophisticated reflecting sextants are in the £200–£300 bracket, and a sextant by Troughton made £280 in early 1973. Sextants made as recently as the 1930s, until recently unregarded, will reach up to £80 in the sale room.

Waywisers and hodometers turn up occasionally in the sale room, and in the 1970–2 seasons their prices varied between £100 and £200; a 1973 sale saw a waywiser realise £620. Theodolites vary greatly in price, ranging from £80 upwards. A Cary theodolite would fetch at least £200, and a Ramsden between £400 and £500. There is not much chance of theodolites being faked, for even at these values it would be difficult to make

121   Twentieth-century sextants, such as this one, will make nearly £100 in the sale room.

122   Good quality eighteenth- and early nineteenth-century instruments are still available at reasonable prices, and pieces such as this circumferentor by J. Coward of Charing Cross are very good investments.

them at the price. Levels are cheaper than theodolites, though the £25 that one would have expected to spend in 1971 for a level has risen to between £50 and £80, and twentieth century levels are snapped up at between £30 and £40.

One of the most collectable of instruments is the telescope. Hand-held telescopes, bought for £10–£15 in 1971, are now worth £30–£35. Refractor telescopes on stands vary enormously in price. There are excellent early nineteenth-century telescopes about for £70 or less, and although one would logically plump for a later nineteenth-century instrument in view of the improvement in optics that characterises telescopes from about 1850 onwards, experts find that early nineteenth-century telescopes are every bit as good as later ones. The greatest improvement has been in coated lenses, a fairly recent innovation.

Microscopes are on a par with telescopes so far as prices go, and fine Victorian instruments with fitted cases can be purchased

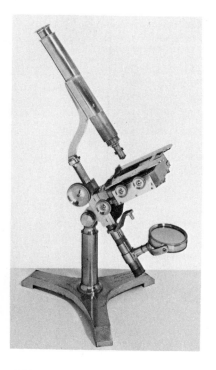

123 Excellent microscopes, such as this one by Ross, are still available at realistic prices.

from £40 and upwards, though eighteenth-century microscopes of the Culpeper type fetch considerably more, but not so much as one would have thought in view of their intrinsic interest. It is interesting that a German 1750 copy of a Culpeper type microscope was sold in 1973 for £240.

Nautical and surveying instruments are closely related, but in nearly all cases those with a nautical cast are worth more than those associated with surveying. Chronometers are highly valued, and the early versions consequent upon their introduction by John Harrison can realise over £2,000. Early nineteenth century chronometers, almost indistinguishable from those of half a century later, realise £300 and more. A Torquay jeweller used a chronometer dated 1864 for twenty-seven years to check watches.

The more specialised instruments, with the exception of early electrical machines, which were more for laboratory and demonstration use than for practical application, are under-collected.

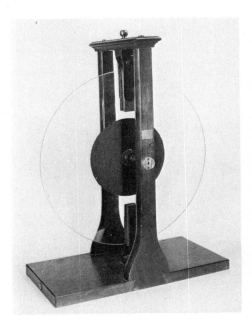

124 An early electrostatic friction machine dated 1799. This made £240 at auction. For more obscure laboratory instruments of the period one would expect to pay much less.

There are few collectors of spectroscopes, for example. Balances vary widely depending on their individuality and usage. Goldsmiths' scales, usually neatly encased with weights, can be obtained for £8 and upwards, and even fine Georgian chemists' scales are in the under £40 class. Brass bankers' scales have jumped in value from £12 or so in 1971 to £35 in 1973.

125 *One would expect to pay a good price for all the instruments depicted on this trade card by John Gilbert, with the exception of the pantograph. Fine Adam-style barometers of the type shown here are very collectable, but perhaps are still undervalued.*

NOTES ON PRICES

Many hopeful collectors have come to grief on barometers. Aneroid barometers, however they are cased, are worth very little. One would expect to pay at least £100 for a good Georgian cistern barometer; stick barometers are decidedly more collectable than those in a heavily moulded frame. Admiral Fitzroy barometers can be bought for £40 or so, though the average price would be much more than this. Thermometers by themselves are not much collected; they are usually incorporated in one unit with a barometer. Barometers with graduated tubes are more sought after than those with a dial.

If they have a provenance, sets of drawing instruments can be worth more than £200. Single instruments, such as pantographs, can be found at very reasonable prices. Finely cased surgical instruments can make surprisingly high prices at auction, proof that this category of instrument has a considerable following.

Scientific instruments as a class have not yet been codified with the precision of, say, porcelain and pot lids. The price one would have to pay for specific items, especially in the provinces, depends very much not on London prices but who is at the sale room and whether there are keen collectors in the neighbourhood. General antique dealers are often not *au fait* with scientific instruments. The owner of a fine early nineteenth-century octant in a shaped case took it to a Bond Street antique shop. He was offered £12 for it. Somewhat disillusioned, he took it to a specialist scientific instrument dealer in Plymouth who immediately offered him £80 for it, an offer that was accepted.

Future events may make the prices laid down in this chapter seem absurd, but there are still bargains to be had in nineteenth-century scientific instruments, and even with eighteenth-century instruments prices are not beyond a sensible level, which is something one cannot say of many categories of antiques.

EIGHTEENTH-CENTURY INSTRUMENT MAKERS

MANY EIGHTEENTH-CENTURY makers of scientific instruments are known from their products, and only known from them. Were it not the custom for makers to sign their instruments, many would languish in complete obscurity. Some are known simply by a signature — sometimes just a surname — on an instrument, and others are known by trade cards or advertisements. A state of affairs exists, therefore, which would be inconceivable in English literature, where the most obscure of writers of the period have been resurrected and put on record. However, one can find an analogy in other antiques. After rattling off Chippendale and Sheraton and a few others, not many people would be able to put names to other period cabinetmakers. In 1800 the situation was improved by the introduction of the Post Office Directory, and in 1877 Kelly's Directory was first produced, providing valuable data about manufacturers etc.

The reason, of course, for the obscurity of eighteenth-century instrument makers is that they were 'trade' and no one thought them worthy of perpetuation in records. The ragtag and bobtail of the aristrocracy who dabbled in philosophical amusements and bought instruments of the Dollonds and the Shorts are recorded with sycophantic comprehensiveness, but the instrument makers had no Almanack de Gotha and their identities have to be sifted from their products and ephemera of the period.

126 Many instrument makers produced a wide range of instruments though it is unusual to find a magic lantern depicted on a trade card, as on this one.

127 *Dudley Adams was fortunate in that he had royal patronage, but most instrument makers languished in obscurity.*

EIGHTEENTH-CENTURY INSTRUMENT MAKERS

BRITISH

Adams, Dudley fl c1795–c1830
Adams, George 1704–73
Adams, George 1750–95
Adams, Nathaniel fl 1737
Alan, J.
Arnold, John fl 1780
Ashley, Thomas fl 1749
Ayscough, James fl 1732–c62

Bakewell, R.
Bancks, fl 1800
Barnett, Thomas
Bass, George fl 1733–69
Bennett, John
Bird, John 1709–76
Bleuler, John 1757–1829
Blunt, Thomas fl 1774–1822
Bolton, Thomas
Bradderley
Brandreth, Timothy fl 1710
Brown fl 1799

Carver, Isaac & Jacob fl c1700
Cary, John fl 1820
Cary, William 1759–1825
Champneys, James
Chaplain
Chapman, J.
Clark, J. (of Edinburgh)
Coggs, John fl 1759
Cole, Benjamin fl c1740–c82
Crichton
Cuff, John 1708–1772
Culpeper, Edmund c1660–1730
Cushee, Richard fl 1760

Dancer, Michael
Davenport, Stephen fl 1720
David, I.
Deane, William fl 1690
Dixey, C. W. fl 1777
Dollond, John 1706–61
Dollond, Peter 1730–1820

Eccleston

Fairbone, Charles fl 1780
Field c1800

Finney, Joseph (of Liverpool) fl 1770–1826
Fowler, John fl 1720–39

Garner, William fl 1734
Gearing, Richard fl 1725
Gilkerson fl c1770
Girl, Martin
Glynn, Richard fl 1710–21
Gould, C. fl c1800
Graham, George 1673–1751
Gregory, Henry fl 1750–92

Haddon, William fl 1713
Hadley, John 1682–1744
Harris, George fl 1733
Harris, R. (of Guernsey) fl 1745
Harrison, John 1693–1776
Hauksbee, Francis fl 1704–15
Hauksbee, Francis (nephew) fl c1725
Hearne, George
Heath, Thomas fl 1729
Herschel, William 1738–1822
Hill, Nathaniel fl 1746–64
Hindley, Henry (of York) 1701–71
Howard, Charles
Hurt, J.

Jackson, Joseph fl 1730–70
Johnson, Samuel fl 1724–72
Jones, David
Jones, Edward
Jones, John
Jones, Thomas fl 1760
Jones, William & Samuel fl 1793–1831

Kimbel, Isaac

Lindsay, George fl 1776
Linnell, Joseph fl 1740–64
Loft, Matthew fl 1711–c47
Long, James

Macall fl 1770
Macy, Benjamin
Mann, James c1660–1730
Mann, James fl c1750

248

EIGHTEENTH-CENTURY INSTRUMENT MAKERS

Marsden, Thomas
Marshall, John 1663–1725
Martin, Benjamin 1704–82
Martin, Joshua fl 1780–c82
Merchant, Robert fl 1745
Mudge, Thomas 1715–94

Nairne, Edward 1726–1806
Newman, James
Newton c1800

Page, Thomas (of Norwich) fl 1735
Parker, William
Patrick, John
Porter, Henry
Price fl 1718
Prijean, John (of Oxford) fl 1667–1701
Pyefinch, Henry fl 1739–90

Quare, Daniel fl 1700

Rabelon fl 1719
Ramsden, Jesse 1735–1800
Reynaldson
Rickett, W. fl 1786
Rowley, John fl 1713

Sanders, Samuel fl 1740–83
Scarlett, Edward fl 1700–43
Scatliff, Samuel fl 1725
Scott, Benjamin fl 1725
Senex, John fl 1728–49

Short, James 1710–68
Short, Thomas fl 1768–?
Shuttleworth, Henry fl 1746–1800
Silberrad
Simms, F. W., 1793–1860
Sisson, Jeremiah fl 1747–c88
Sisson, Jonathan 1690–1749
Smith, Addison fl 1764–74
Stedman, Christopher fl c1780
Sterrop, George
Streatfield

Troughton, Edward 1753–1835
Troughton, John fl 1770–84
Tulley, Charles fl 1782–1830
Tuttell, Thomas fl c1700

Urings, J. fl 1752

Watkins, Francis fl 1737–c74
Watkins, J. & W. fl 1791–c1805
Watkins, Thomas fl 1746
Watson, James
West, Charles
Wilson, Alex fl 1740
Wilson, James c1665–1730
Wing, Tycho
Winter, Thomas fl 1800
Worgan
Wright, Thomas 1711–86
Wynne, Henry fl 1710–21
Yarwell, John fl 1700

AMERICAN

Bailey, John fl 1778
Bailey, John, II 1752–1823
Baily, Joel (practitioner) 1732–97
Baldwin, Jedidiah c1777–1829
Banneker, Benjamin (practitioner) c1734–1806
Benson, John fl 1793–97
Biddle, Owen (practitioner) 1737–99
Biggs, Thomas fl 1792–95
Blakslee, Ziba 1768–1834
Blundy, Charles fl 1753
Bowles, Thomas S. c1765–1821
Breed, Aaron 1791–1861
Brokaw, Isaac fl 1771
Balmain & Dennies fl 1799
Burges, Bartholomew fl 1789
Bernap, Daniel 1759–1838
Caritat, H. fl 1799
Chandlee, Benjamin Jr. 1723–1791
Chandlee & Bros. fl 1790–91
Chandlee, Ellis 1755–1816
Chandlee, Ellis & Bros. fl 1791–97
Chandlee, Goldsmith c1751–1821
Chandlee, Isaac 1760–1813
Clark, Robert fl 1785
Clough, Jere 18th century
Condy, Benjamin fl 1756–98, d 1798
Crow, George c1726–72
Dabney, John, Jr. fl 1739
Dakin, Jonathan fl 1745
Davenport, William 1778–1829
Dean, William (?–1797)
Devacht, Joseph and Francois fl 1792
Donegan (or Denegan), John fl 1787
Donegany, John (see Donegan)
Doolittle, Ernos 1751–1806
Doolittle, Isaac 1721–1800
Doolittle, Isaac, Jr. 1759–1821
Dupee, John fl after 1761
Ellicott, Andrew (also practitioner) 1754–1820
Emery, Samuel 1787–1882
Evans, George fl 1796; d 1798
Fairman, Gideon (See Hooker and Fairman) 1774–1827
Fisher, Martin fl 1790
Folger, Peter (practitioner) 1617–90
Folger, Walter, Jr. 1765–1849
Ford, George fl late 18th century to 1842
Ford, George, II fl 1842
Fosbrook, W. fl 1786 or earlier

EIGHTEENTH-CENTURY INSTRUMENT MAKERS

Gatty, Joseph   fl 1794
Gilman, Benjamin C.   1763–1835
Gilmur, Bryan   fl end of 18th century
Godfrey, Thomas   1704–49
Gould, John   fl 1794
Grainger, Samuel (practitioner)   fl 1719
Greenleaf, Stephen   1704–95
Greenough, Thomas   1710–85
Greenough, William   fl 1785
Greenwood, Isaac, Sr. (practitioner)   fl 1726
Greenwood, Isaac, Jr.   1730–1803
Grew, Theophilus (practitioner)   fl 1753
Hagger, Benjamin King   c1769–1834
Hagger, William Guyse   c1744–1830?
Halsie, James, I. (practitioner)   fl 1674
Halsy, James, II   1695–1767
Halsy, John   fl 1700
Halsy, Joseph   fl 1697–1762
Ham, James   fl 1754–64
Ham, James, Jr.   fl 1780
Hamlin, William   1772–1869
Hanks, Benjamin   1755–1824
Hanks, Truman   fl 1808
Harland, Thomas   1735–1807
Heisely, Frederick A.   1759–1839
Heisely, George   1789–1880
Hinton, William   fl 1772
Hoff, George   1740–1816
Holcomb, Amasa (also practitioner)   1787–1875
Hooker & Fairman (William Hooker and Gideon Fairman)   before 1810
Houghton, Rowland   c1678–1744
Huntington, Gurdon   1763–1804
Jacks, James   fl 1780s
Jayne, John   late 18th century
Kennard, John   1782–1861
Ketterer, Alloysius   fl 1789
King & Hagger (Benjamin King and William Guyse Hagger)   1760s
King, Benjamin   1707–86
King, Benjamin   1740–1804
King, Daniel   1704–90
King, Samuel   1748–1819
Lamb, A. & Son   1780s
Lamb, Anthony   1703–84
Lamb, John   1735–1800
Mendenhall, Thomas   fl 1775
Miller, Aaron   fl 1748–71
Morris, M.   fl 1785
Newell, Andrew   1749–98
Newell, Joseph   fl 1800–13
Pease, Paul   fl 1750
Platt, Augustus   1793–1886

251

Platt, Benjamin   1757–1833
Pope, Joseph   1750–1826
Potter, John   fl 1746–1818
Potts, W. L.   late 18th century
Prince, John (practitioner)   1751–1836
Prince, Nathan (practitioner)   fl 1743
Pryor, Thomas   fl 1778
Revere, Paul   1735–1818
Rittenhouse, Benjamin   1740–c1820
Rittenhouse, David (practitioner)   1732–96
Rittenhouse & Evans   fl 1770s
Sibley & Marble (Clark Sibley and Simeon Marble)   late 18th century
Smith, Cordial   fl 1775
Sommer, widow Balthaser   fl 1753
Sower, Christopher   c1724–40
Stiles & Baldwin (Jedidiah Baldwin)   fl 1791
Stiles & Storrs (Nathan Storrs and Jedidiah Baldwin)   fl 1792
Taws, Charles   fl 1795
Thacher, Charles   18th century
Thaxter, Samuel   1769–1842
Voight, Henry   1738–1814
Wall, George, Jr.   fl 1788
Walpole, Charles   fl 1746
Warren, Benjamin   fl 1740–90
White, Peregrine   1747–1834
Whitney, John   fl 1801
Whitney, Thomas   fl 1798–1823
Williams, William   1737 or 1738–92
Willis, Arthur (practitioner)   fl 1674
Wilson, James   1763–1855
Wistar, Richard   fl 1752
Witt, Christopher (practitioner)   fl 1710–65
Wood, John   fl 1790
Youle, James   1740–86
Youle, John   fl 1786

SCIENTIFIC-INSTRUMENT MAKERS OF LONDON

1843

BAROMETER and THERMOMETER MAKERS
†Philosophical instrument makers
*Looking-glass makers
§Opticians
‡Hydrometer makers
‖Saccharometer makers

*Battistessa & Co, 106 Hatton Gardens
Brugger, Lorenz & A., 79 High Holborn
†Calderara, Serafino, 78 Leather Lane
Carrughi, Paul, 128 High Holborn
Clark, Robert, 27 Brooke St, Holborn
Colomba, A., 37 Charles St, Hatton Garden St E, Boro'
‖Couldrey, Jph., 26 St Thomas St, E Boro'
Dring & Fage, 20 Tooley Street
Fagioli, Dominic & Son, 3 Gt Warner St
Gally, P. P. & Co, 50 Exmouth St Spitalfields
Johnson, William, 34 Hatton Garden
Maltwood, Rd. A., 19 Charles St, Hatton Garden, & 3 Orange Row, Kennington Rd
Martinelli, Alf., 96 Vauxhall St, Lambeth
†Palmer, Edward, 103 Newgate St
Pastorelli, A., 4 Cross St, Hatton Garden
Pensa, Mrs Margt, 25 Chas. St, Hatton Garden
‡‖Ronketti, John, 15 Museum St, Bloomsbury
§Somalvico, Joseph & Co, 2 Hatton Garden
Tagliabue & Casella, 23 Hatton Garden
†Tagliabue, Angelo, 91 Leather Lane
†Tagliabue, Anth, 31 Brooke St, Holborn
†Tagliabue, John, 11 Brooke St, Holborn
Ward, John, 79 Bishopsgate Within

CHEMICAL-APPARATUS MAKERS
Doulton & Watts, 28 High St, Lambeth
Green, Stephen, Princes St, Lambeth
Knight, George & Sons, 2 Foster Lane
Palmer, Edward, 103 Newgate St
Ward, John, 79 Bishopsgate Within

HYDROMETER and SACCHAROMETER MAKERS
Bedwell, Thos, 53 Gt. Alie St, Goodm filds
Dring & Fage, 10 Duke St, Tooley St
Long, Joseph, 20 Little Tower St
Ronketti, John, 15 Museum St
Somalvico, Joseph & Co, 2 Hatton Garden

MATHEMATICAL-INSTRUMENT MAKERS
†Opticians
*Drawing-instrument makers
§Philosophical-instrument makers
‡Nautical-instrument makers
*Allen, John, 35 St Swithin's Lane

SCIENTIFIC-INSTRUMENT MAKERS OF LONDON

§†Barry, Geo & Chas, 4 Luke St, Finsbury
†Bate, Robert Brettell, 21 Poultry
Beale, Wm Thos, 10 King St, Whitechapel
‡Carrew, John W., 13 Wapping Wall
†§Clarke, Edward M., 428 Strand
Claxton & Morton, 27 Harrington St N
‡Cole, Thos, 21 Hannibal Rd, Stepney Green
Couldrey, Jsph, 26 St Thomas's St, Boro'
Coulsell, Wm., 9 Castle St, Boro'
*Coulson, Dan., 58 Charles St, City Rd
†Crichton, John, 112 Leadenall St
Cuthbert, Chas., 9 Clerkenwell Green
†Davis, John, 38 New Bond St
Dennis, John C., 118 Bishopsgate Within
Derry, Chas., 7 Leigh St, Burton Crescent
*Drake, J. C., 19 Elm St, St John's Wood
†*§Elliott, William, 268 High Holborn
Endersbee, Wm & Son, 28 Wapping
Ettling, Leopold, 151 Ratcliffe Highway
Evans, Wm, 48 Rosamon St, Clerkenwell
Fayrer, J. & Son, 66 White Lion St, Pentonville
Fleming, John, 41 Brick Lane, Spitalfields
§Gogerty, Robert, 32 King St, Smithfield
§Grafton, Henry, 80 Chancery Lane
Green, William, 14 Fountain Pl, City Rd
†*§Harris, Wm, & Son, 50 High Holborn
Hobcraft, Wm, 14 Barbican
Huddy Francis, 37 Duke St, Smithfield
†Hughes, Henry (*whol.*), 120 Fenchurch St
†‡Hughes, Joseph, 38 Queen St, Ratcliff
Lilley, John, 71 Jamaica Ter, Com. R. Lime
†*Macrae, Henry, 34 Aldgate
Mills, George, 82 Parsons St, Ratcliff
*Neeves, William, 67 High St, St Giles's
Nelson Henry, 2 Glo'ster Ct, Queen Square

‡*§Newman, John, 122 Regent St
Oborne, B., 11 Guildford St E, Spitalfields
Pallant, John, 14 Mercer St, Long Acre
†§Palmer, Edward, 103 Newgate St
Parsons, James, 50 Red Lion St, Clerkenwell
Penney, William, 25 Broad St, Ratcliff
Piggott, Peter Wm, 4 Penton St, Walworth
Piggott, Wm P., 11 Wardrobe Pl, Doc. Com.
Price, Henry, 58 Brook St, Lambeth
Price, William, 115 Fetter Lane
Rooker, John, 1, Lit. Queen St, Holborn
†§Schmalcalder, John, 2 Fairfax's Court, 400 Strand
Scott, James Geo., 17 Bermondsey Wall
Simms, Jas. & G., Greville St, Hatton Garden
Simpson & Irwins', 54 Hatton Garden
Slemmon, Thos., 3 Brandon Row, New Kt. Rd
Snart, Miss Neariah, 35 King St, Boro'
Soulby, J. & Co., New Stairs, 126 Wapping
†Spencer, Browning & Co., 111 Minories
§Stiles, Wm., 29 Seward St, Goswell St
Syeds, Mrs. Agnes, 379 Rotherhithe
Thompson, Joseph Berry, 36 Wapping
†Troughton & Simms, 138 Fleet Street
§Ward, John, 79 Bishopsgate Within
§Wood, Henry & Co, 1 Long Lane, Smithfield
Wrench, Edward, 6 Gray's Inn Ter
‡§Yeates, Andrew, 12 Brighton Pl, N. Kent Rd

NAUTICAL-INSTRUMENT MAKERS

Blachford & Imray, 116 Minories
Carrew, John Wm., 13 Wapping Wall
Cole, Thos, 21 Hannibal Rd, Mile End
Crichton, John, 112 Leadenall St
Hughes, Henry, 120 Fenchurch St
Hughes, Joseph, 38 Queen St, Ratcliff
Jeacock, James, 32 Fore St, Limehouse
Syeds, Mrs. Agnes, 379 Rotherhithe
Taylor, George, 103 Minories
Whitbread, G., 2 Grenada Terr, Com. Rd E

OPTICAL TURNERS

Bruce, Wm., 16 King's Hd, Ct, Shoe Lane
Gunston, Mich, 6 St James's Wk, Clerkenwell
Marriott, Wm, 38 Montague St, Spitalfields

OPTICIANS

†Mathematical- and philosophical-instrument makers
*Drawing-instrument makers
§Hydrometer makers
‖Saccharometer makers
‡Barometer and thermometer makers

Armadio, Francesco, jun, 35 Moorgate St
Amadio, Francesco, sen, 118 St John St Rd
Amadio, J., 6 Shorter's Ct, Throgmorton St
Ansell, Joseph, 80 Leman St, Goodman Fields
†§Bate, Robert Brettell, 21 Poultry
Bithray, Steph, 6 Spread-eagle Ct, Finch. 1
Bracher, Mrs. M., 19 King St, Commercial Rd, E
Bradfield, Wm, Hen, 31 Royal St, Lambeth
Bradford, George, 99 Minories
Bruton, Jas, 25 Pleasant Row, Pentonville
†*Carpenter & Westley, 24 Regent St
†Cary, William, 181 Strand
Catmur, Benj, 28 Chamber St, Goodman Fields
Childe, Henry, 66 Vauxhall Walk, Lambeth
†Chislett, Alf, 8 Postern Row, Tower Hill
Clarke, Edward Marmaduke, 428 Strand
Clift, Charles, 3 New Inn Yd, Shoreditch
Comyns, Heny, 17 King's Rd, Chelsea
†Cox, James, 5 Barbican
Crawley, Wm, 21 Oxenden St, Haymarket
Cushion, Daniel, 2 New Compton St, Soho
Davenport, Chas, 33 Tabernacle Row
†Dixey, Chas. Wastell, 3 New Bond St
†Dixey, Edward, 335 Oxford St
Dixey, William, 241 Oxford St
Dobson, John, 54 Newington Causeway
†Dollond, Geo, 59 St Paul's Churchyard
Dowling, George Wm, 16 Mortimer St
†*Elliott, William, 268 High Holborn
Fairey, J. & Son, 8 Northumberland Pl, Com Rd E
Frith Brothers (J. Greatbach, agent), 46 Lisle St & Arundel S,, Sheffield
Froggatt, Saml, 13 Charterhouse St
Gilbert, William, 138 Fenchurch St
Harris & Son, 52 Gt Russell St, Bloomsbury
Hawes, William, 95 Cheapside
Hemsley, Henry, 138 Ratcliff Highway
†Hemsley, Thos, 11 King St, Tower Hill
†Hillum, Richd, 109 Bishopsgate Within
Hogg, William, 394 Rotherhithe St
Holmes, John & Wm, 14 Redcross Sq
Horton, George M. & Co, 36 Ludgate St
Houliston, James, 33 New Bond St
Hughes, Henry, 120 Fenchurch St
†Hughes, Joseph, 16 Queen St, Ratcliff
Hynes, Abraham, 35 Gt Prescot St
Irvine, Mrs Elizh, 32 Kirby St, Hatton Garden
James, Wm, 25 Little Russell St, Bloomsbury
Jeffreys, William, 26 Wilderness Row
Johnson, Wm, 188 Tottenham Ct Rd
†Jones, Wm & Samuel, 30 Holborn Hill
Jones, James, 25 Eyre St Hill
†Jones, James William, 87 Goswell St
†*§Jones, Thomas, 62 Charing Cross
Keohan, Thos, Arbour Ter, Commer. Rd E
Littlewort, Geo, 11 Ball All, Lombard St
Littlewort, Wm., 7 Ball Alley, Lombard St
Lowe, Albert, 18 Warwick Sq, Newgate
Marratt, John, 15 Gt Winchester St
Matthews, T. B., 9 Athol Pl, Pentonville
Millard, Joseph & Son, 24 Coppice Row
Moon & Fleming, 76 Minories
Munro, James, 4 High St, Lambeth
Murrell, John, 13 Albion Pl, St John's Lane
Ottway, John, 11 Devonshire St, Queen Sq
†Palmer, Edward. 103 Newgate St
Parker, James, 39 Theobald's Rd
Parnell, Thomas, 2 Low East Smithfield

SCIENTIFIC-INSTRUMENT MAKERS OF LONDON

Parry, Thos Wm, 24 Holywell St, Strand
Phelps, Mrs Mary, 5 City Terr, Old St Rd
Phillips, Solomon, 231 Tottenham Ct Rd
Pilkington, Geo, 14 Clarence Pl, Pentonville
†*§‖‡Pizzala, A., 7 Charles St, Hatton Garden
Porter, Jas, 8 Brownlow St, Drury Lane
†Pritchard, Andrew, 162 Fleet St
Pye, Wm, 20 Albion Bdgs, Bartholomew Cl
*†Rennoldson, I, 5 Cambridge Pl, Hackney
Ripley, J., 15 Warkworth Ter, Comml Rd E
†§‖Ronketti, J. G. H., 116 Gt Russell St, Bloomsbury
†*§‖‡Ross, Andrew, 21 Featherstone Bdgs
Rowley, John, 26 Edgware Rd
†Rubergall, Thomas, 24 Coventry St
Salmon, Wm John, 105 Fenchurch St
Samuel, E. I. & Co, 73 Prescot St
Sargent, A., 17 Ryder's Court, Leicester Sq
†*§‖‡Schmalcalder, John, 400 Strand
Smith, Joseph, 15 Palace Row, New Rd
Solomons, Sam & Benj, 39 Albermarle St, Piccadilly, & 76 King William St, City
Solomons, Elias, 36 Old Bond St
§†Somalvico, Jsph & Co, 2 Hatton Garden
Spencer, Browning & Co, 111 Minories
Street, Thos, 39 Commercial Rd, Lambeth
Sullivan, Chris, 22 Charles St, Hatton Garden
Sutton, George, 16 Bridge Rd, Lambeth
Tolley, Ed, 8 Tavistock Row, Covent Garden
†Walker, Francis, 35 Wapping Wall

SURGICAL INSTRUMENT MAKERS

Amesbury, Joseph, 8 Berners St Oxford St
Armitage, John, 2 Hampton St, Walworth
Arnold, Jas & John, 35 West Smithfield
Bailey, Wm Huntley, 418 Oxford St
Ballard, Hen, 42 St John Sq, Clerkenwell
Bartrop, Henry, 33 Crawford St
Blackwell, Wm, 3 Bedford St, Covent Garden
Brennand, Pearson, 217 High Holborn
Coxeter, J. & Co, 23 Gratton St E, Tottenham Ct Rd
Craddock, George, 35 Leicester Sq
Dixon, Wm, 2 Tonbridge St, New Rd
Einsle, Edward, 46 St Martin's Lane
Elam, Alfred & Bros, 403 Oxford St
Ellis Wm (*glass cylinder*), 3 Thanet Pl, Temple Bar
Evans, John & Co, 10 Old Change
Everill & Co, 67 St James's St
Evrard, Jean, 35 Charles St, Middlesex Hosp
Ferguson, Daniel, 21 Giltspur St
Figgett, J. L., 29H, Trafalgar St, Walworth
Fisher, James, 7 Cannon St Rd
Fuller, John, 239 Whitechapel Rd
Grumbridge, J. (*stethos.*), 42 Poland St
Hemming, Augs, F., 45 Chiswell St, Finsbury
Hentsch, Fredk, C., 25 Bartletts Bldgs
Higham, Mrs Priscilla, 48 Jermyn St
Hughes, Francis & Co, 247 High Holborn
Hunter, Mason & Co, 44 Webberrow, Blackfriars
Jack, W., 16 Cloth Fair, West Smithfield
Johnson, Henry, 121 Blackfriars Rd
Kemp, John, 7 New St, Cloth Fair
Laundy, Jsph, 9 St Thomas's St, Boro'
Lindsey, Mark, 18 Webb St, Borough
Lings, William, 1 Jewin St
Maw & Stevenson, 11 Aldersgate St
Millikin, John, 301 Strand
Mundy, Wm, 6 Finsbury St, Chiswell St
Nye, Joseph, 10 Bridge House Pl, Boro'
Nye, Samuel, 16 St Andrew's Rd, Borough
Paget,, William, 195 Piccadilly
Paul, John Adam, 13 Up St, Martin's Lane
Pepys, William H., 22 Poultry
Philip & Whicker, 67 St James's St
Pryor, Thomas, 23 King St, Tower Hill
Read, John, 35 Regent Circus, Piccadilly
Rein, F. Charles (*acoustic*), 340 Strand
Savigny & Co, 67 St. James's St
Settle, Dennis, 9 High St, Marylebone

SCIENTIFIC-INSTRUMENT MAKERS OF LONDON

Shoolbred & Renwick, 34 Jermyn St
Simpson, Mrs Susan, 55 Strand
Smith, Samuel, 40 Gray's Inn Lane
Smith, W., 2 New St, St Thomas's St, Borough
Sparks & Co, 28 Conduit St, Regent St
Stevens, Jas, 10 Gower St N, New Rd
Thompson, J., 38 Gt Windmill St, Haymarket
Ward, F. W., 3 Athol Pl, New Rd, Pentonville
Weedon, Thos, 41 Hart St, Bloomsbury
Weiss, John & Son, 62 Strand, & King William St, Strand
Whicker & Philip, 67 St James's St
Whitfield, William, 104 High Holborn
Whiting, William, 15 King St, Boro'
Willson, William, 22 Woodbridge St, Clerkenwell
Wright, Henry, 18 London Rd, Southwark

1894

BAROMETER and THERMOMETER MAKERS

†Philosophical-instrument makers
‡Hydrometer makers
§Opticians
∥Saccharometer makers
¶Aneroid-barometer makers

Adie, Patrick, Broadway Wks, Westminster St
Angell, Wm Geo, Earl's Bdgs, Featherstone St EC
¶Barker, F. & Son, 12 Clerkenwell Rd EC, speciality—pocket & surveying aneroids
Bendon, Geo & Co, 36 & 37 Ely Pl EC & 1 Charterhouse St EC
§†Browning, John, 63 Strand WC
∥‡Calderara, Serafino & Alfred, 10 Cross St, Hatton Garden EC
†Callieu & Bartlett, 11A Featherstone Bldgs WC
‡∥Cetti, Edward, 36 Brooke St, Holborn EC
Chick, Edwd, 24 Woodbridge St, Clerkenwell EC
‡Comitti, O. & Son, 69 Mount Pleasant, Gray's Inn Rd WC; patentees of the "Visible" aneroid dial barometers & sole manufacturers of the Torricelli barometer
¶Daniels, John, 14 Smith St, Clerkenwell EC
Denton, Samuel Geo, 25A Hatton Garden EC
†§Dollond & Co, 35 Ludgate Hill EC; 62 Old Broad St EC & 5 Northumberland Av WC
∥§†Doublet, T. & H., 11 Moorgate St EC
Eagland, Joseph, 152 Farringdon Rd WC
Eidmans, Joseph S. & Co, 15 Bartlett's Bldgs EC
‡∥†Griffin, John J. & Sons Limited, 22 Garrick St WC & 2 Long Acre WC
†Harvey & Peak, 56 Charing Cross Rd WC
Hawkes, John, 2 Albany St, Camberwell SE
¶§Heath & Co Limited, 115 & 117 Cannon St EC; works, Crayford, Kent
†Hicks, James J., 8, 9 & 10 Hatton Garden EC & St Peter's House, Clerkenwell Rd EC
Hughes, Henry & Son, 59 Fenchurch St EC
†‡∥¶§Levi, Joseph & Co (*whol.*), 40 Furnival St EC & 2 Dyers' Bldgs, Holborn EC
Levi, S. J. & Co (*whol.*), 71 Farringdon Rd EC
Lilley, John & Son, 10 London St EC
Lynch & Co Limited, 192 Aldersgate St EC
McCarthy, William Henry, 9 Mount Pleasant WC
†‡§∥Mark, J. & Co, 76 Fann St EC
Marlow, Harry, 5 Mount Pleasant WC
Marson, Francis, 32 Clerkenwell Green EC
†¶Negretti & Zambra, 38 Holborn Viaduct EC; branches, 45 Cornhill EC & 122 Regent St W
‡∥Owers & Willson, 51 East Rd N
†§‡∥¶Pastorelli & Rapkin (*whol.*), 46 Hatton Garden EC
¶Pastorelli, Alfred, 13 Featherstone Bldgs WC
Penso, Victor, 55 Hatton Garden EC & 8 & 13 Hatton Yard EC

†§¶Perken, Son & Rayment (*whol.*), 99 Hatton Garden EC. Trade mark, Optimus
†Peters Wm, 11 Spencer St, Goswell Rd EC
¶§†Pillischer, Jacob, 88 New Bond St W
†‡¶Pitkin, James (*whol.*), 56 Red Lion St, Clerkenwell EC; manufacturers of pocket watch & other aneroid barometers, hydrometers, electrical & philosophical instruments
‡∥Pizzala, Chas F., 26 Charles St Hatton Garden EC
Rawling, William, 85 Hatton Garden EC
¶Reynolds, T. A. & Co, 15 Sekforde St EC
‡Ronketti, Joseph, 25 Northampton Rd EC
¶∥‡Short & Mason, 40 Hatton Garden EC
Simmons, Geo Alexander, 90 Hatton Garden EC
§Société des Lunetiers (*whol.*), 56 Hatton Garden EC
†¶§Steward, James Henry, 406 & 457 Strand WC & 7 Gracechurch St EC
Tyler, John, 3 Jerusalem Bldgs, Jerusalem Court, Clerkenwell EC
¶Wheeler, Thos, 8 Coombs St, City Rd N
Wilson, R. H. C., see Pastorelli & Rapkin
‡Witherspoon & Rudd, 11 Fox Court, Gray's Inn Rd EC

CHEMICAL GLASS and APPARATUS MANUFACTURERS

Baird & Tatlock, 14 Cross St, Hatton Garden EC
Becker, F. E. & Co, 33, 35 & 37 Hatton Wall EC
Brayne, Francis & Co, Three Mills Lane, Bromley-by-Bow E
Breffit, Edgar & Co Limited, 83 Up Thames St EC
Caplatzi, Anthony, 3 Chenies St WC & 22 Charlotte St W
Clarke, Daniel George, 48 Brunswick Close EC
Doulton & Watts, 28 High St, Lambeth SE & Albert Embankment SE; manufacturers of condensing worms, acid pumps, retorts, receivers, mixing pans, stearine pans, store jars, &c, warranted to resist the strongest acids. Numerous Prize Medals.
Ford, John & Co, 20 Bartlett's Bldgs EC
Gallenkamp, Adolf & Co, 2, 4 & 6 Cross St, Finsbury EC
Griffin, John J. & Sons Ltd, 22 Garrick St WC & 2 Long Acre WC
Horne & Thornthwaite, 416 Strand WC
Hughes, William, all kinds of chemical apparatus made to order, 175 Bow Rd E
Linley, Owen, 12A Dean St, Holborn WC
May, Roberts & Co, 9 & 11 Clerkenwell Rd EC
Orme, John & Co, 65 Barbican EC
Poths H. & Co (importers), 4 Creechurch Lane EC
Powell, Samuel, 37 Swan St, Gt Dover St SE
Rouch, William White & Co, 180 Strand WC & 43 Norfolk St, Strand WC
Smith, Thomas & Co, receivers, worms acid bottles &c, Canal potteries SE; sample rooms, 50 & 51 Lime St EC
Stiff, J. & Sons, London Pottery, Lambeth SE; Retorts, receivers, store jars, taps, plumbago & clay crucibles, porous cells, cylinders, plates, stoneware insulators & all other pottery for electric & chemical work
Townson & Mercer, 89 Bishopsgate St, Within EC & 11 Bishopsgate Avenue EC

ELECTRICAL- and GALVANIC-INSTRUMENT MAKERS

Baird & Tatlock, 14 Cross St, Hatton Garden EC
Brown, Ferdinand William, 25 Charlotte St, Caledonian Rd N
Casella, Louis P., 147 Holborn Bars EC
Coxeter, James & Son, new patented dry cells for Voltaic & Faradic batteries & for lighting; low price; absolute constancy & reliability, small space & light weight. Prize medals & highest awards 1851 to 1892 & Gold medal, 4 & 6 Grafton St, Gower St WC
Darlow & Co, 89 New Oxford St WC
Elliott Brothers, 101 & 102 St Martin's Lane WC

SCIENTIFIC-INSTRUMENT MAKERS OF LONDON

Griffin, John J. & Sons Ltd, 22 Garrick St & 2 Long Acre, Covent Garden WC
Harvey & Peak, 56 Charing Cross Rd WC
Hughes, Henry & Son, 59 Fenchurch St EC
India Rubber, Gutta Percha & Telegraph Works Co Ltd, 100 & 106 Cannon St EC
International (The) Electric Co; sole depôt of Mix & Genest, Ltd, 55 Redcross St EC
Johnson & Phillips, 14 Union Court, Old Broad St EC
Koerber, John Adam, 44 Frith St, Shaftesbury Ave W
Mudford, Francis John, Jackson Rd, Holloway N
Muirhead & Co, 2 Princes St; works, Princes Mews SW
Negretti & Zambra, 38 Holborn Viaduct EC; branches, 45 Cornhill EC & 122 Regent St W
Oakley, Waltr Hy & Co, 202 & 203 Grange Rd SE
Orme, John & Co, 65 Barbican EC
Perken, Son & Rayment (*whol.*), 99 Hatton Garden EC
Pillischer, Jacob, 88 New Bond St W
Pitkin James, 56 Red Lion St, Clerkenwell EC; makers of ammeters, voltmeters & recording instruments, &c
Pulvermacher & Co, 194 Regent St W
Schall, Karl, 55 Wigmore St W
Société des Lunetiers (*whol.*), 56 Hatton Garden EC
Townson & Mercer, 89 Bishopsgate St, Within EC & 11 Bishopsgate Ave EC
Walters, P. & Co, 249 & 251 Kensal Rd W

ELECTRICAL MEASURING-INSTRUMENT MAKERS

Electrical Co Ltd (The), 122 & 124 Charing Cross Rd WC
Hodges & Todd, 82 Turnmill St EC
Nalder Bros & Co, 16 Red Lion St, Clerkenwell EC

HYDROMETER and SACCHAROMETER MAKERS

See also Barometer & Thermometer Makers; also Opticians.
Burrow, Walter & John, 62 & 63 Gt Tower St EC

Buss, Thomas Odempsey (metal; three prize awards), 33 Hatton Garden EC
Calderara, Serafino & Alfred, 10 Cross St, Hatton Garden EC
Casella, Louis P., 147 Holborn Bars EC
Farrow & Jackson, 16 Gt Tower St EC; 8 Haymarket SW; & factory, 89 & 91 Mansell St E
Lilley, John & Son, 10 London St EC
Loftus, William Robert, 321 Oxford St W & 2 & 3 Lancashire Court, New Bond St W
Long, Joseph, 43 Eastcheap EC
Macord & Arch, 47 Gt Tower St EC
Negretti & Zambra, 38 Holborn Viaduct EC; branches, 45 Cornhill EC & 122 Regent St W
Oertling Ludwig, Turnmill St EC
Pastorelli & Rapkin, 46 Hatton Garden EC
Potter, John Dennett, 31 Poultry EC & 11 King St, Tower Hill E

MATHEMATICAL-INSTRUMENT MAKERS

†Optician
*Drawing-instrument maker
§Philosophical-instrument makers
‡Nautical-instrument makers
Adie, Patrick, Broadway Works, Westminster St SW
§Archbutt, William Edwards, 201 Westminster Bridge Rd SE
†*Aston & Mander, 25 Compton St W
Barker, F. & Son, wholesale makers of ship, pocket & prismatic compasses, sextants, marine & pocket aneroids, drawing & surveying instruments, Baker's patent altitude instruments, &c, 12 Clerkenwell Rd EC
Bennett, Sir John, Ltd, 65 Cheapside EC
Boulter, Cornelius Alfd, 7½ St John's Lane, Clerkenwell EC
†§‡Browning, Louis P., 147 Holborn Bars EC
Clarkson Alex, 28 Bartlett's Bldgs, Holborn Circus EC; wholesale manufacturers of surveying instruments
*Cooper, Chas Thos, Russell Pl, 9 Old Kent Rd SE
†§‡Dollond & Co, 35 Ludgate Hill EC; 62 Old Broad St EC & 5 Northumberland Ave WC
*‡§†Elliott Bros, 101 & 102 St Martin's Lane WC
Evans, Chas Jas, 20 Silver St, Bloomsbury WC

*Eyre & Spottiswoode, Gt New St, Fetter Lane EC
*Finch, Mrs Eliza, 30 Ranwell St, Bow E
Foulon & Quantin (Paris) (G. H. Saunders, agent), 37 Farringdon St EC
George, Robert, 652 Commercial Rd East E
Green, Wm, 5 Mount Pleasant, Clerkenwell WC
†Gregory, Wm & Co, 51 Strand WC
*Harling, Wm Hy, 47 Finsbury Pavement EC
§Hardt, Wm, 165 Pentonville Rd N
‡§Holmes, F. (level), 76 Shacklewell Lane E
Holtzapffel & Co, 64 Charing Cross SW & 127 Long Acre WC
§§*†Horne & Thornthwaite, 416 Strand WC
†*‡Hughes, Henry & Son, 59 Fenchurch St EC
Lack, Phillip (agent for J. Schröder, Darmstadt), 10 Wood St Sq EC
Liddon, James & Frederick, 28 Wynyatt St EC
†§‡Lilley, John & Son, 10 London St EC
†*‡§Negretti & Zambra, 38 Holborn Viaduct EC; branches, 45 Cornhill EC & 122 Regent St W
Newton, Henry, 63 Bancroft Rd E
*Nicholl, Robert, 153 High Holborn WC
†§‡Pastorelli & Rapkin (whol.), 46 Hatton Garden EC
§Pearce, Geo Ed, 22A New Church St, Strand WC
Peover, Wm, 27 Leigh St, Burton Cres WC
†§‡Perken, Son & Rayment (whol.), 99 Hatton Garden EC
†§‡Pillischer, Jacob, 88 New Bond St W
*Reeves & Sons Ltd, 113 Cheapside EC
*‡†§Ross & Co, 111 New Bond St W
*Scott, Robert, 262 Goswell Rd EC
Short, John, 2 Gladstone St, London Rd SE
*Sloper, Alfred & Son, 4A Rockingham St, Newington Causeway SE
‡Société des Lunetiers (whol.), 56 Hatton Garden EC
*Stanley, Wm Ford, 4 & 5 Gt Turnstile WC
*†‡Steward, James Henry, 406 & 457 Strand WC & 7 Gracechurch St EC
Stone, Wm, 41 Gloucester St WC
Tree, James & Co, 7 Lawrence Lane EC
†Troughton & Simms, 138 Fleet St EC
*‡†Watson Bros, 31 Cockspur St SW
†*Watson, W. & Sons, 313 High Holborn WC; & steam factory, 9, 10 & 11 Fullwood's Rnts WC
Watts, Edwin Richard, 123 Camberwell Rd SE
†§Wood, Edward George, 74 Cheapside EC

MICROSCOPE MANUFACTURERS

See also Opticians.

Cary, Wm & Co, 181 Strand WC
Clarkson, Alexander, 28 Bartlett's Bldgs, Holborn Circus EC; wholesale manufacturers of monocular & binocular microscopes
Elliott Bros, 101 & 102 St Martin's Lane WC
Gregory, Wm & Co, 51 Strand WC
Negretti & Zambra, 38 Holborn Viaduct EC; branches, 45 Cornhill EC & 122 Regent St W
Newton, Frederic & Co, 3 Fleet St EC
Perken, Son & Rayment (whol.), 99 Hatton Garden EC
Société des Lunetiers (whol.), 56 Hatton Garden EC
Steward, Jas Hy, 406 & 457 Strand WC & 7 Gracechurch St EC
Watson, W. & Sons, 313 High Holborn WC; steam factories, 9, 10, 11, 16 & 17 Fullwood's Rents WC; wholesale & retail. 34 gold & other medals at international exhibitions, Paris, Chicago, London, &c
Weeden, Chas Cartwright, 15 Merlin's Pl, Clerkenwell EC
Weeden, Edward John, 68 Myddelton Sq EC
White, George, 16 Furnival St, Holborn WC

NAUTICAL-INSTRUMENT MAKERS

*Compass makers
Atkins, George Wm, 144 Lower Rd SE
*Barker, F. & Son, 12 Clerkenwell Rd EC. See mathematical instrument makers
Barton, Thos, 21 Maroon St, Limehouse Fields E

SCIENTIFIC-INSTRUMENT MAKERS OF LONDON

*Butters, Geo Wm, 4 Cannon St Rd E
Cannon, Wm Freeman, 175 Shadwell High St E
Cary, Wm & Co, 181 Strand WC
Cooper & Wigzell, Billiter House, Billiter St EC; sea sounding apparatus makers
*Dent, E. & Co, 61 Strand WC & 4 Royal Exchange, Cornhill EC; factory, 4 Hanway Pl W; makers of Admiralty compasses
Eidmans, Jsph S. & Co, 15 Bartlett's Bldgs EC
*Elliott Bros, 101 & 102 St Martin's Lane WC
Elliott, Thomas, 7 Belgrave St, Stepney E
George, Robert, 652 Commercial Rd East E
*Heath & Co Ltd, 115 & 117 Cannon St EC; works, Crayford, Kent
Hemsley, Thomas & Son, 4 Tower Hill E & 4 King St, Tower Hill E
*Herbert, Thomas, 146 New North Rd N
*Hughes Henry & Son, 59 Fenchurch St EC
James' Syndicate Limited (sounding machines), 18 Billiter St EC
Kelvin, Lord, compass & sounding machine; (sole London agents, John Lilley & Son), 10 London St, Fenchurch St EC
*Levi, Joseph & Co (*whol.*), 40 Furnival St EC & 2 Dyer's Bldgs, Holborn EC
*Lilley, John & Son, 10 London St EC
McGregor & Co, 14A London St EC
*Negretti & Zambra, 38 Holborn Viaduct EC; branches, 45 Cornhill EC & 122 Regent St W
*Oliver, Geo & Jsph, 37, 38 & 39 Wapping Wall E
*Perken, Son & Rayment (*whol.*), 99 Hatton Garden EC
Potter, John Dennett, 31 Poultry EC & 11 King St, Tower Hill E
*Reynolds & Son, 32 Crutchedfriars EC & 7 Dionis Yard EC
*Ritchie, E. S. & Son (Negretti & Zambra, sole agents), 38 Holborn Viaduct EC
Sargent, Thos C. (*whol.*), 37 Odessa St, Rotherhithe SE
*Short & Mason, 40 Hatton Garden EC
*Steward, James Henry, 406 & 457 Strand WC & 7 Gracechurch St EC

Walker, T. & Son, 58 Oxford St, Birmingham; makers of ship's logs, sounding machines, compass needles & bar magnets
Watson Brothers, 31 Cockspur St SW
*Wiggins, F. & Sons, 10 Tower Hill E & 102 Minories EC
Wilson, Geo, 23 Sherwood St, Piccadilly Circus W

OPTICIANS

† Mathematical- and philosophical-instrument makers
* Drawing-instrument makers
§ Hydrometer makers
‖ Saccharometer makers
‡ Barometer &c. makers
¶ Spectacles makers

¶Ackland, William (surgeon), 416 Strand WC
Adie, Patrick, Broadway Works, Westminster SW
Airy, Geo (*who*), 42 Gray's Inn Rd WC
¶‡*†Apps, Alfred, 433 West Strand WC
†Archbutt, William Edwards, 201 Westminster Bridge Rd SE
Archer, Reuben, 17 Myddelton St EC
*‖‡¶Baker, Charles, 244 High Holborn WC
Barker, John & Co, 71 Kensington High St W
Barrett, Charles, 144 Fleet St EC
¶Beck, R. & J., 68 Cornhill EC & Holloway Rd N; & Lister Works, Dickenson St. Kentish Town NW
Bendon, Geo & Co. (*whol.*), 36 & 37 Ely Place EC & 1 Charterhouse St. EC
Bluett, Fredk. John, 8A Great Portland St. W
¶Botwright & Grey, 13 Spencer St, Clerkenwell EC; repairs & orders by post executed the same day
Braham Octavius (*whol.*), 3 Sekforde St EC
British & Foreign Optical Co Limited, 24 Bartlett's Buildings EC
†Browning, John, 63 Strand WC
‡Bruce, John Lewis (*whol.*), 56 Clerkenwell Rd EC
Callaghan, William & Co, 23A New Bond St & 69 Beak St W
*†§‖‡Caplatzi, Anthony, 3 Chenies St Bedford Sq. WC & 22 Charlotte St

¶†*Carpenter & Westley, 24 Regent St SW
Cary, William & Co. 181 Strand WC
Comitti, O. & Son (*whol.*), 69 Mount Pleasant WC
Conway, Joseph, 2 Tysoe St, Clerkenwell EC
Coppock, Chas (oculists'), F.R.A.S. 26 Maddock St W
Cox, Frederick, 98 Newgate St EC
†*‡¶Crouch, Henry Limited, 66 Barbican EC
Culver, Geo (manftg.), 99 to 105 White Lion St N
Curry & Paxton (late Pickard & Curry), 195 Great Portland St W
Dallmeyer, J. H. Limited, 25 Newman St W
†§‖‡Darton F. & Co (*whol.*), 142 St John St EC
¶‡†Dixey, Charles W. & Son, 3 New Bond St W
Dixey, William & Son, 552 Oxford St W
†*‡¶Dollond & Co, 35 Ludgate Hill EC; 62 Old Broad St EC & 5 Northumberland Ave WC
‡*†¶Doublet, T. & H. 11 Moorgate St EC
†*§‖‡¶Dring & Fage, 145 Strand WC
Druiff, Lionel, 44 Hatton Garden EC
†*¶Eidmans, Jsph S. & Co 15 Bartlett's Bldgs EC
Elkan, John, 35 Liverpool St EC
¶‡*†Elliott Bros, 101 & 102 St. Martin's Lane WC
Ellis, Geo Everest, 63 King William St EC
Finnemore, Jabez, 31 Skinner St., Clerkenwell EC
†Gotz, Jn Rodolphe, 19 Buckingham St, Strand WC
Gould & Porter, 181 Strand WC
Gray, Charles, 11 Crooked Lane, Cannon St EC
Gray-Keith, J. 44 Queen Victoria St EC
Greenwood, Edward, 22 Northolme Rd N
‡*†¶Gregory, William & Co. 51 Strand WC
To Her Majesty's Government, War Department, London County Council & National Rifle Association by appointment, also to over thirty County, Indian & Colonial Rifle Associations & principal Rifle Clubs in Great Britain. 28 years at Wimbledon camp; 4 years at Bisley
¶Grünfeld, Maurice Moritz 62 & 63 St John's Sq EC
Hancock, John, 61 City Rd, Finsbury EC
¶Harris, Thos. & Son, 32 Gracechurch St, EC originally estd. opposite British Museum 1780
Harrison, Thos. Harnett, 40 Hatton Garden EC
†‖¶Harvey & Peak, 56 Charing Cross Rd WC
Hawes, Alfred, 79 Leadenhall St EC & 7 Great Marylebone St W
‡†Heath & Co Limited, 115 & 117 Cannon St EC; Works, Crayford Kent
Henri, T. & Co, 34 Aldersgate St EC
Henri, Thomas, 43 Victoria St SW
Heydon, George, 9 Cross St, Hatton Garden EC
†‡Hicks, James J. 8, 9 & 10 Hatton Garden EC & St Peter's House, Clerkenwell Rd EC
Hilger Adam, 204 Stanhope St NW
¶*‖†Horne & Thornthwaite, 416 Strand WC
Horwitz, Daniel & Co. 20 High Holborn WC
Howes & Isaac (*whol.*), 85 Hatton Garden EC
†§‖‡¶Hudson & Son, Greenwich SE; to the Admiralty, R.N. College & Royal Observatory
Hudson, A. C. & Co, 5 Crosby Sq EC
Hughes, Henry & Son, 59 Fenchurch St EC
Hughes, Wm Chas, Brewster House, Mortimer Rd, Kingsland N; magic lantern manufacturer
Hummel, Maurice F. (*whol.*), 23 Fenchurch St EC
‡¶Isaacs M. L. (*whol.*), see Joseph Levi & Co
†‡¶Jacquemin (J.B.) Bros, wholesale manufrs. of optical goods of all kinds, 65 Hatton garden EC; manufactory, Morez du Jura, France; Paris depôt, 23 Rue Béranger
Johnson, John Severin, 1 & 1A, Lower Charles St Northampton Sq EC
¶Johnson, William, 188 Tottenham Court Rd W

SCIENTIFIC-INSTRUMENT MAKERS OF LONDON

Jones, Samuel, 294 Walworth Rd SE
Jones, William, 55 Myddelton St EC
Kemp, Henry, 7 Thavie's Inn EC
Kimbell & Cole, Aldgate Ave, Aldgate High St E
Krauss, E. & Co (*whol. manfg.*), 3 Paper St EC
Lamb & Co. (*whol. & export*), 6 Benyon Rd N
Larkins, George Hick, 37 Ardleigh Rd N
¶Laurance, Henry, 44 Hatton Garden EC
Lawley, Walter, 78 Farringdon St EC
Lawrence & May, 67 & 69 Chancery Lane EC
¶Le Claire, August, 67 Praed St W
∥Lenton & Rusby, Waingate, Sheffield
Leon, L. K. & Co, 167 Piccadilly W
*∥‡§†¶Levi, Joseph & Co (*whol.*), 40 Furnival St EC & 2 Dyers' buildings, Holborn EC wholesale agents for Richard's (of Paris) celebrated recording instruments
¶Levi, S. J. & Co (*whol.*), 71 Farringdon Rd EC
†*§∥‡¶Lilley, John & Son, 10 London St EC
London & Paris Optic & Clock Co, 24 Edmund Place EC & 7 Jewin St EC
†London Stereoscopic & Photographic Co. Ltd 54 Cheapside EC & 106 & 108 Regent St W
Lorberg, Charles Hy, 38 Kensington High St W & 6 Station Buildings, South Kensington W
Luby Brothers, 113 Regent St W
Lucas Joseph, 59 King's Rd Chelsea SW
Mackinney, Fredk Williams, 156 High St, Notting Hill W & 263 Liverpool Rd, Islington N
†§∥‡¶Mark J. & Co, 76 Fann St EC
¶Marratt & Ellis, 63 King William St EC
¶Martin, Robert & Son (*whol.*), 17 Albion Buildings, Aldersgate St EC
Martin, George Sallnow, 3 Wigmore St W
Messer, George Bracher, 80 Christian St, St. George's East E
Millard, Thomas & Son, 245 Oxford St W
Millard, John 60 Upper St, Islington N
Monnier Bros (*whol.*), 48 Hatton Garden EC

Murrell Chas Hy, 1 Irnmnger rw, St Luke's EC
Nalder Bros. & Co, 16 Red Lion St Clerkenwell Rd EC
National "Silex" Optical Co (The), 138 & 138A, Strand WC & 2 Princes Buildings, Coventry St W
†*§∥‡¶Negretti & Zambra, 38 Holborn Viaduct EC; branches, 45 Cornhill EC; 122 Regent St W & Crystal Palace SE
Newbold, Alfred Edwin, 149 Goswell Rd EC
Newbold, Wm M., 36 Charles St., Hatton Garden EC
*¶†Newton, Frederic & Co, 3 Fleet St EC
‡¶Nitsche & Günther (*whol.*), 66 Hatton Garden EC
Norman, Charles, 119 Old St EC
Norris, Geo. Alfd, 13 New Chas. St, City Rd EC
Oakley, Walter Henry & Co (apparatus), 202 & 203 Grange Rd SE
Optical Institute of London, 94 Hatton Garden EC
Ottway, John & Son, 178 St John St Rd EC
Pastorelli, Frank & Co. 10 New Bond St W
‡∥§†Pastorelli & Rapkin (*whol.*) 46 Hatton Garden EC
Peacock Charles Gilbert, 425 Strand WC
Pearce, Henry, 52 Park St, Camden Town NW
Pearce, John, 2 Windmill St, Tottenham Ct Rd, W
Pearce, Stanley, 161 Wardour St, Soho W
Percy & Royle, 24 Wilton Rd SW
†‡¶Perken, Son & Rayment (*whol.*), 99 Hatton Garden EC & 141 Oxford St W; Trade Mark "Optimus"
Pickard, E. S. & Co., 83 Great Portland St W
¶Piggott, John, 117 Cheapside EC & 1 & 2 Milk St Buildings EC
∥†‡¶Pillischer, Jacob, 88 Bond St W; medals 1851, 1855, 1862, 1873, 1878, the decoration of the Imperial Francis Joseph Order; gold medal, Paris, 1889
¶‡*†Porter, Henry, 181 Strand WC
Powell & Lealand, 170 Euston Rd NW
Preist & Co, 514 Oxford St W

263

‡¶Priest & Ashmore, Newton Works, Earl St, Sheffield
Raphael J. & Co (*whol.*), 13 Oxford St W
Richard & Co., 24 Cannon St EC
Robbins, George & Co, 4 Brewer St W
Robinson Edward Henry & Co, 52 Bishopsgate Within EC
Robinson, J. & Sons, 172A, Regent St W
†*‡¶Ross & Co. 111 New Bond St W; gold medals, London, 1851 & 1862, Paris 1867 & Philadelphia, 1876, Paris, 1878 & Inventions Exhibition, 1885; grand prix & gold medal Paris, 1889; Kingston, Jamaica 1891, award, Chicago 1893, for first class microscopes, telescopes & photographic lenses & apparatus, race glasses &c
Rowley, John & Son, 60 Edgware Rd W
Rowley, William Nelson & Co, 6 Mount Pleasant, Gray's Inn Rd WC
Salmon, John & Son, 169 Hampstead Rd NW
†‡¶Salt, William, 65 Hatton Garden EC
Samuels, Albert, 61 East St, Manchester Sq W
¶Sharland, Herbert Henry (*whol.*), 7 & 8 Thavie's Inn EC
Simmons, E. & R. (*whol.*), 57 Red Lion St EC
‡¶Simmons & Fredericks (*whol.*), 19 Charterhouse Buildings EC
Slade, John & Co, 17 Mile End Rd E
Slape, Harraway, 22 Camden Rd NW
Smith, Henry & Son, 16 Park Side SW
†Societé des Lunetiers, 56 Hatton Garden EC; & 6 rue Pastourelle, Paris (wholesale) optical, electrical, nautical & mathematical instruments, spectacles, barometers, opera glasses, telescopes, microscopes, time & hour glasses, electric bells, rules, magic lanterns, lenses & slides, stereoscopes, steroscopic views, photographic apparatus & photographic lenses
Somalvico Joseph & Co, 16 Charles St, Hatton Garden EC
Spencer, Wm James, 167 Gray's Inn Rd WC
¶Spiller Geo, 3 Wigmore St., Cavendish Sq W
Springate, R. & Co, 295 Oxford St W
¶Stanbury, William, 12 Charterhouse St EC
†*Stanley, Wm Ford, 13 Railway Approach, London Bridge SE
†*‡¶Steward, James Henry, optician to The British & Foreign Governments, the Board of Education, South Kensington, the National Rifle Associations of England, Ireland, India, Canada & America, the National Artillery Association, by appointment, 406 & 457 Strand WC & 7 Gracechurch St EC; maker of the celebrated Lord Bury telescope; maker of the Fitzroy barometers, as in use at all railway termini & principal hotels in London.
¶‡Sutton Geo & Son, 209 Westminster Bridge Rd SE
†‡¶Sutton, Charles Thos, 108 Holloway Rd N
Swift, James & Son, 81 Tottenham Court Rd W
Theobald John & Co, 20 Church St, Kensington W & (*whol.*) 43 Farringdon Rd EC
Travers, Kessell & Co, 50 Hatton Garden EC
Walter, Berger & Co (*whol.*), 21 Hatton Garden EC
¶‡*†Watson W. & Sons, 313 High Holborn WC; & steam factories, 9, 10, 11, 16 & 17 Fullwood's Rents, Holborn WC.
†*‡¶Watson Brothers, 31 Cockspur St SW
Webster Brothers, 4 Porchester Rd W
¶Wells & Lyon, 7 Myddelton St EC
Whitehouse, Joseph, 37 Warwick St, Pimlico SW
Wiggins, F. & Sons, 10 Tower Hill E & 102 Minories E
Wilson, Robert, 15 Merchant St, Bow Rd E
Wilson Wm, 56 Crogsland Rd Chalk Farm NW
*Wise, Edward Thomas, 554 Old Ford Rd, Bow E
Wood, late Abraham, 20 Lord St., Liverpool

SCIENTIFIC-INSTRUMENT MAKERS OF LONDON

‡¶†Wood Edward George, 74 Cheapside EC
¶†Wood Henry Joseph, 185 Oxford St W
Wrench John & Son (*whol.*), 50 Gray's Inn Rd WC; wholesale opticians, magic lantern & slide manufacturers

SURGICAL-INSTRUMENT MAKERS

See also Acoustic Instrument Makers; also Bougie & Catheter Makers; also Orthopædic Instrument Makers.

* Working surgical instrument makers
† Veterinary instrument makers
‡ Truss makers

*†‡Arnold & Sons, by appointment to Her Majesty's government &c., &c. (trusses, elastic stockings, belts, artificial legs, arms, &c. Veterinary instrument manufacturers, by appointment to the Royal Veterinary college), 31 West Smithfield & 1, 2 & 3 Giltspur Street EC (estblished 1819)
‡Bailey, W. H. & Son, 38 Oxford St W
*‡Baker, Charles, 243 High Holborn WC
Beauchamp, Rt Jas, 40 & 41 West Smithfield EC
Bell Henry, Guildford St York Rd SE & 154 York Rd, Lambeth SE (india rubber only). Enamas, injection bottles, balls, pessaries, pads, urinals; india rubber articles of every description made to order
Bourjeaurd, Philip, 22 Davies St, Berkeley Sq W
Brittain, George & Co, 9 Little Britain EC
Brooker, Frederick (elastic), 56 Methley St SE
Burge, Warren & Ridgley, 11 Clerkenwell green EC; manufacturers of every description of hard & soft india rubber surgical instruments & appliances, also in ivory, bone & wood; vulcanite in sheets, rods & tubes of various sizes & thicknesses or made to order
*Carsberg Geo & Son, 8 Meredith St, Clerkenwell EC
Carter, Alfred, 47 Holborn Viaduct EC; warehouse, 49 Farringdon St EC; operating tables & gynæcological couches
Chapman, Thos, 21 Gloucester St, Clerkenwell EC
‡Chapman, Wm Henry & Sons, 8, 9 & 10 Stock Orchard St, Caledonian Rd N
Chemists' (The) Assoctn Ltd, Curtain Rd EC
Coleman, Wm, 10 Brewer St, Goswell Rd EC
Collier, A. & Co, 21 St James' Walk EC
Collins, Daniel J., 21 Poland St, Oxford St W
Cow (P. B.) & Co. 46 & 47 Cheapside EC & 66 Bread St EC; works, Streatham common SW; air & water beds, hospital sheets, india rubber bandages, film gloves &c
‡Coxeter, James & Son, 4 & 6 Grafton St WC
Dansey James, 42 & 43 Ufford St, Lambeth SE
Domen Belts Warehouse, 61 Moor Lane EC
*‡Down Bros, 5 & 7 St Thomas St, Boro' SE; manufactory, King's Head inn yard, 45 Borough High St SE
Durst, Thomas, 38 Poland St W
*†‡Eggington, Thomas, 7 & 9 Grosvenor St, Manchester
Elges, W. 49 Barbican EC
Eschmann Bros. & Walsh, 23 Bartholomew Sq EC
*†‡Evans & Wormull, 31 Stamford St SE by appointment to the Army, Navy & Indian Government
Gimber Walter, 2 Linden Villas, Harder's Rd, Peckham SE
Goercke, Fredk, 82 Bevenden St, Hoxton N
Goldschmidt, Alfred, 49 Barbican EC
†‡Gray, Joseph & Son, Truss Works, Boston St, Sheffield. See advertisement.
Hague Joseph John, 75 Pentonville Rd N
‡Hawksley, Thos, 357 Oxford St W & Sedley Pl W
‡†Hedgcock & Co, 143 Camberwell Rd SE
Hentsch, Frederick, 49 Greek St, Soho W
Hill, Zachariah, 1 Ann St, Islington N
Hills & Co. 46 Newcomen St, Borough SE

SCIENTIFIC-INSTRUMENT MAKERS OF LONDON

Hillyard, Thomas, 88 Horseferry Rd SW
Hutchinson, A. & Co. 70 Basinghall St EC; air & water beds, hospital sheeting, india rubber bandages, air bed (registered) for camping out
†*Hutchinson, W. & H. Matilda St, Sheffield. See advertisement
Huxley, Edward, 13 Old Cavendish St, W
Innocent, Thomas Henry, 4 Canonbury Rd N
*Jack Robert & Son, 200 Goswell Rd EC
*†‡Krohne & Sesemann, 8 & 38 Duke St Manchester Sq W; 241 Whitechapel Rd E & 14 Barrett St W
Lambert E. & Son, manufacturer of silver & electro-plated surgeons' instruments & druggists' sundries, 60 & 62 Queen's Rd, Dalston NE
Lange Hermann Julius, Alma Grove, Copenhagen St, Caledonian Rd N & 47 Arthur Rd, Tollington Rd, Holloway N
Lawley Walter, 78 Farringdon St EC
*†Lee J. & Co (late Lee & Lambert) 45 Matilda St Sheffield
‡Lindsey & Sons, 32 Ludgate Hill EC & 40 Gracechurch St EC
Ludski, Barnet & Son, 63 Commercial St E
†‡Lynch & Co. Limited, 192 Aldersgate St EC
Mabon, John, 36 Charles St, Hatton Garden EC
McMillan, John H. 1 Rahere St, Goswell Rd EC
*Manson Richard, 75c Fortess Rd NW
Mascall Brothers, 1 Weedington Rd NW
‡Mather William, 94 Milton St EC
Matthews Brothers, 10 New Oxford St WC
Maw, S, Son & Thompson, 7 to 12 Aldersgate St EC
May, Roberts & Co, 9 & 11 Clerkenwell Rd EC
Mayer & Meltzer, 71 Great Portland St W
Medical Supply Assoc, 96 High Holborn WC
Metzeler & Co., 49 Barbican EC
Millikin & Lawley, 165 Strand WC
Mohr & Co, 57 Frith St, Soho Sq W

Müller, C. 64 Holborn viaduct EC; works, 89 Neue König Strasse, Berlin, N.O.; vulcanite, india rubber & metal goods
New York Hamburg India Rubber Co (The), 8 Coleman St EC
*Nicholls, Thomas & Son, 258 Kingsland Rd NE
Nyman, Henry, 33 Charles St, Hatton Garden EC
Perren, Wm Henry, 1A Bowling Green Lane EC
Powell & Barstow, 58 Blackfriars Rd SE
Powell John Francis & Son, 13 Englefield Rd N
†Read & Co, 139 Oxford St W
*Robinson, Arthr & Prcy, 3 Helmt St, 338 Strand WC
Sanders, Francis Henry, 23 Osnaburgh St NW
Schall Karl, 55 Wigmore St W
Schollar & Simsky, 64 Praed St, Paddington W
‡Schutze, Fredk. & Co, 36A Aldersgate St EC
Shrimpton (Eml.) & Fletcher, Première works, Redditch; needles, instrument accessories, ligatures &c; established 1810
Skull Theodore (appliances), 91 Shaftesbury Av W
Spratt W. H. & Brooke, 48 New Bond St W
Stacy, Wm. King, 4 Newgate St EC; chatelaines, wallets & instruments for nurses
‡†Stevens James & Son, 78 Long Lane EC
Still & Co, 20 Charles St, Hatton Garden EC; specialities; ear trumpets, metal splints, slings, specula, inhalers, bed, hot air & mercurial baths, bronchitis kettles (various), blood pans, brass labels & general tin & copper work
‡Strudwick, Charles, 57 & 56 Up. Marlebone St W
Tytheridge & Hummel, 94 Rosoman St EC
Walsh, John (gum elastic), 158 Cobourg Rd SE
Walsh Brothers (elastic gum), 4 Caroline Place, Marlborough Rd SW
‡Walters, Frederick & Co. 29 Moorgate St EC & 69 Lambeth Palace Rd SE
Ward, Henry & John, 95 Kentish Town Rd NW

Ward, Horace & Co, 350 Upper St N
Warne, Wm & Co, 29 Gresham St EC; works, Tottenham, Middlesex; india rubber syringes, catheters & the new patent "Safety Filling" water bottle &c.
Watson, James, 25 Richmond St, St Luke's EC
†‡Weiss, John & Son, 287 Oxford St W (formerly, 62 Strand)
Weller, Frederick William, 20 Argyle St WC
Willis, Hy Benj, 185 Euston Rd NW
Wood, Vincent, 3 St Andrew St EC
‖Wright, C. & Co., from Louis Blaise & Co. (late Savigny & Co.), 108 New Bond St W

SURVEYING-INSTRUMENT MAKERS

Adie, Patrick, Broadway wks, Westminster SW
Barker F. & Son, 12 Clerkenwell Rd EC See Mathematical Instrument Makers
Cary, William & Co, 181 Strand WC
Casella, Louis P., 147 Holborn Bars EC
Couling, F. & Co, Garnault Mews, Rosebery Ave EC
Elliott Bros, 101 & 102 St Martin's Lane WC
Esdaile & Son, 28 Queen's Row, Walworth SE
Gent, Frank E., 96 Rosoman St, Clerkenwell EC
Groves Rchd, Philip, St Alban's Pl, Islington N
Hall Brothers, 53 Spencer St EC
Hicks, James J. 8, 9 & 10 Hatton Garden EC & St. Peter's House, Clerkenwell Rd EC
Negretti & Zambra, 38 Holborn Viaduct EC; branches, 45 Cornhill EC & 122 Regent St W
Steward, James Henry, 406 & 457 Strand WC & 7 Gracechurch St EC

Watson, W. & Sons, 313 High Holborn WC; & steam factories, 9, 10, 11, 16 & 17 Fullwood's Rents, Holborn WC

TELESCOPE MAKERS
See also Opticians

Cary, William & Co, 181 Strand WC
Clarkson, Alex, 28 Bartlett's buildings Holborn circus EC; wholesale manufacturer of astronomical telescopes with high class object glasses; maker to the trade of the $3\frac{1}{2}$ feet astronomical & day telescope with 3 in. object glass on stand, suitable for astronomy, sea views, rifle ranges &c. marine, naval & tourist telescopes; any specification carried out.
Dallmeyer, J. H. Limited, 25 Newman St W
Dollond & Co, 35 Ludgate Hill EC; 62 Old Broad St EC & 5 Northumberland Ave WC
Eidmans, Jsph, S. & Co., 15 Bartlett's Bldgs EC
Elliott Brothers, 101 & 102 St. Martin's Lane WC
Hammersley, Joseph, 41 Halfmoon Crescent N
Horn & Thornthwaite, 416 Strand WC
Lilley, John & Son, 10 London St EC
Negretti & Zambra, 38 Holborn Viaduct EC; branches, 45 Cornhill EC & 122 Regent St W
Newton, Frederic & Co, 3 Fleet St EC
Ross & Co, 111 New Bond St W
Société des Lunetiers (*whol.*), 56 Hatton Garden EC
Steward, James Henry, 406 & 457 Strand WC & 7 Gracechurch St EC
Watson, W. Sons, 313 High Holborn WC; steam factories, 9, 10, 11, 16 & 17 Fullwood's Rents, Holborn WC
Wood Edward George, 74 Cheapside EC

BIBLIOGRAPHY

I HAVE DELIBERATELY restricted myself to books in the English language in this bibliography. It is too often assumed that all those interested in scientific instruments are polyglots, and for those readers who wish to get to grips with works in a foreign language I would refer them to the admirable bibliography at the end of Francis Maddison's article on scientific instruments in Volume 5 of the *Connoisseur's* encyclopaedia of antiques. The omission of works in a foreign language emphasises how ill-served collectors of scientific instruments are, though it is gratifying to note that the more authoritative works in German or French are being translated into English, such as Maurice Daumas' *Les instruments scientifiques aux XVIIe et XVIIIe siècles,* published in 1972 in its English form.

One work omitted from the bibliography but which is worthy of special mention is the *Encyclopaedia Britannica,* particularly the eleventh edition, which although more than fifty years old deals with its subjects with admirable precision and decisiveness. For those fascinated by technical and mathematical detail in surveying and navigation there are no better source books.

Special attention should be drawn to the mammoth Oxford *History of Technology,* in five volumes to 1900.

Adams, George.   *Geometrical and Graphical Essays* (1791)

Bannister, A. and Raymond S.   *Surveying* (1965)

Bedini, Silvio A.   *Early American Scientific Instruments* (Washington 1964)

BIBLIOGRAPHY

Bell, G. H. and E. F.   *Old English Barometers* (1952)
Bergen, W. C.   *Practice and Theory of Navigation* (1872)
Bernal, J. D.   *Science and Industry in the Nineteenth Century* (1953)
Bolton, H. C.   *Evolution of the Thermometer* (Easton 1900)
Bradbury, Savile.   *Evolution of the Microscope* (1967)
Brewington, M. V.   *Peabody Museum Collection of Navigating Instruments* (Salem 1963)
British Optical Institute.   *Dictionary of British Scientific Instruments* (1921)

Calvert, H. R.   *Globes, Orreries and Other Models* (1967)
Chaldecott, J. A.   *Temperature Measurement and Control* (1955)
Clay, R. S. and Court, T. H.   *History of the Microscope* (1932)
Clerke, A. M.   *History of Astronomy during the Nineteenth Century*, 4 vols (1903)
Close, Sir Charles.   *The Early Years of the Ordnance Survey* (1926)
Cotter, C. H.   *History of Nautical Astronomy* (1968)
Cousins, Frank W.   *Sundials* (1969)

Daumas, Maurice.   *Scientific Instruments of the Seventeenth and Eighteenth Centuries* (1972)

Goodson, N.   *English Barometers 1680–1860* (1969)
Gore, J. H.   *Geodesy* (Boston and New York 1891)
Gould, R. T.   *The Marine Chronometer* (1925)
Grant, R.   *History of Physical Astronomy* (1852)
Griffiths, E.   *Methods of Measuring Temperature* (1925)
Gunther, R. T.   *Early Science in Cambridge* (1937)
———   *Early Science in Oxford* (1923–45)

Harris, John.   *Lexicon Technicum*, 2 vols (1710)
Hewson, J. B.   *History of the Practice of Navigation* (Glasgow 1951)
Hill, H. O. and Paget-Tomlinson, E. W.   *Instruments of Navigation* (1958)

BIBLIOGRAPHY

Hitchens, H. L. and May, W. E. *From Lodestone to Gyro Compass* (1955)

Hogg, J. *The Microscope,* 15th edition (1898)

Inskip, R. M. *Navigation and Nautical Astronomy* (1865)

Jeans, H. W. *Navigation and Nautical Astronomy* (1858)

Josten, C. H. *Catalogue of Scientific Instruments from the Thirteenth to the Nineteenth Century from Collection of J. A. Billmeir* (1955)

Kiely, E. R. *Surveying Instruments* (New York 1947)

King, Henry. *History of the Telescope* (1956)

Love, John. *Geodasia* (1688)

Martin, Benjamin. *New Elements of Optics* (1759)

Mayall, R. N. and Mayall, M. L. *Sundials* (Boston 1938)

Middleton, W. E. K. *History of the Barometer* (Baltimore 1964)

Moxon, J. *Mechanick Dyalling* (1703)

Mudge, W. *An Account of the Trigonometrical Survey of England and Wales,* 3 vols. (1801–11)

Palmer, F. W. and Sahiar, A. B. *Microscopes to the End of the Nineteenth Century* (1971)

Ramsey, L. G. G. *Concise Encyclopaedia of Antiques* compiled by the *Connoisseur* (1961)

Rees, A. *Cyclopaedia* (1819)

Richeson, A. W. *English Land Measuring to* 1800 (1966)

Roe, J. W. *English and American Tool Builders* (New Haven 1916)

Routledge, Robert. *Discoveries and Inventions of the Nineteenth Century* (1900 edition)

Singer, Charles, Holmyard, E. J., Hall, A. R. and Williams, T. I. *A History of Technology,* vols 4 and 5 (1958)

Smiles, Samuel. *Industrial Biography* (1879)

Stanley, W. F. *Mathematical Drawing and Measuring Instruments* (1888)

Stanley, W. F. *Surveying and Levelling Instruments* (1901)
Stevenson, E. L. *Terrestial and Celestial Globes* (1921)
Stock, J. T. *Development of the Chemical Balance* (1969)
Stone, Edmund. *Construction and Principle Uses of Mathematical Instruments* (1723)

Taylor, E. G. R. *Mathematical Practitioners of Hanoverian England* (1966)
——— *Mathematical Practitioners of Tudor and Stuart England* (1954)
Thoday, A. G. *Astronomical Telescopes* (1971)
Tomlinson, C. *Cyclopaedia of Useful Arts* (1852-4)

Wartnaby, J. *Surveying* (1968)
Waters, D. W. *The Art of Navigation in England in Elizabethan and Early Stuart Times* (1958)
Wright, A. E. *Principles of Microscopy* (1906)

In addition there are articles contained in technical and learned journals such as *Nature,* the *Transactions* of the Newcomen Society, the *Journal of the Institute of Navigation, Mariner's Mirror, Mechanic's Magazine, English Mechanic, Invention, Philosophical Magazine,* the *Philosophical Transactions* of the Royal Society, *Engineer* and *Expositor.* There is also interesting material to be found in Victorian magazines, such as *Cassell's Magazine, All the Year Round, Household Words,* and the *Illustrated London News,* couched in language acceptable to the general reader.

INDEX

Page references in italics denote illustrations

Abbé, Ernest, 99, 131
achromatic lens, 41–2, 86, 88, 110, 128
achromatic microscope, *129*
actinometer, 200
Adams, George, *17*, 48, 71, *127*, 129, 147
Addison, 59
Adie, 142
Airy, George, 94–5
altazimuth, 38
Alter, David, 110
aluminium, 174–5, 187
aluminium bronze, 174
America, 46–9, 63–4, 93, 95, 107, 131, 140–1
Amici, G.B., 130–1
Amontons, G., 137–8, 147
amputating saw, *166*
anemometer, *201–2*
aneroid barometer, *143–5*, 244
apertometer, *200*
aplanatic lens, 110
apochromatic lens, 131
aquarium microscope, *132*
d'Armati, Salvino, 75, 118
armillary sphere, *72*, 101–3

Arnold, John, 60
artificial horizon, 58, *59*
artillery quadrant, 33
assay balance, 154
astrolabe, 15, *21*, 22, 31, *53*, 54–5
astrolage, 55
astronomical quadrant, 97–8
astronomy, 22, 33, 55, 74–114
Ayscough, James, 87, 128

back staff, 55, 64
Bacon, Roger, 74, 118
balances, 154–8
banker's scales, 242
Barlowe, 52
barograph, *144*
barometer, 18, 90, 134–45, 244
Beckmann, E., 150
Berthoud, F., 61
Berthoud, P.-L., 61
Best, John, *167*
bimetallic thermometer, 150
binocular microscope, 123, 129, 131, *132*
binoculars, 98–9
Bird, John, 40, 86, 98
Blondeau, J. de, 139

INDEX

Borda, J. C. de, 26, 45
Boulanger, A. A., 99
Bourdon, E., 143
bow compasses, 162
box sextant, 57–8
Boyle, Charles, 103
Boyle, Robert, 147
Bramah, Joseph, 178–9
brass, repairs to, 171, 173, 175, 183–5
Breguet, A. L., 61
Brewster, David, 110
Browning, John, 111, 115
Brunel, I. K., 162
Burnap, Daniel, 47
Butterfield, Michael, *69,* 70, 238

Cabot, G., 53–4
Calver, George, 96
Campani, G., 122
Carter, Henry, 64
Cary, William, 90
Cassengrain, 80
Cavendish, Charles, 150
Cavendish, Henry, 156
Cetti & Co, 149
chain, 24–7
Champneys, James, 88, 141
Chance Brothers, 19, 95
Chaplain, 81
charts, *36,* 50, 58
Chaulnes, Duc de, 43
chemists' scales, 242
Chevalier, V. & C., 120, 128
chromatic aberration, 75–6
chronometer, 50, 57–63, 242
circumferentor, 28, *29,* 31–4, 43, *240*
cistern, 138–9, *142*

Clark, Robert, 64
clinical thermometer, 148–51
clock making, 43, 60, 177
clockwork drive, 77
Cock, Christopher, 78
Cole, Benjamin, 49, 105, 107
colorimeter, *209*
Columbus, Christopher, 53
Combs, Oliver, *246*
compass, *26, 35,* 38, 47, 50–4, *209, 211*
compasses, 159, 162
compound microscope, 119–20
compressor, *213*
conical barometer, 137
Conté, Nicholas, 140, 143
Cook, Captain, 58, 60
Cooke, Thomas, 95
Cox, John, 78–9
Cuff, John, 126–9
Culpeper, Edmund, 124–6
Cushee, Richard, *101,* 108
cystoscope, 170

Dabney, John, 63
Daniell, J. F., 153
Davis, John, 55
Dawes, W. R., 95
Deane, William, 71
Descartes, R., 118–19, 135
diagonal barometer, 138
dial barometer, *136*
dials, 67–73, *238–9*
diptych dials, 68–9, *70*
dissecting microscope, *130*
distance finder, *65,* 66
dividers, 162
dividing engine, 40, *41–2,* 43, 180–1

273

INDEX

Divini, 122
Dollond, George, 89, 90, 93
Dollond, John, *30,* 40, *44,* 86–9, *98,* 110, 128, 161
Dollond, Peter, 86, 88–9, 92
Donegany, Joseph, 140
double microscope, *124*
double tripod microscope, *125*
drawing instruments, 159–63, 244
Drebbel, Cornelius, 120
Dring, Thomas, 140, 149
drum microscope, 126
Dumpy level, 41
Dupee, John, 47
dynanometer, *214*

Earnshaw, Thomas, 60
electrolysis, 169
Ellicott, Andrew, 49
Elliott, 73
Ellis, O. W., 47
Elton, John, 59
equatorial telescope, *97*
equinoctial dial, 71–2
Everest theodolite, 46

Fitzroy, Admiral, 140, 244
Flinders, Matthew, 52
Focque, Nicolas, 177
folded barometer, 137
forceps, 166
Fortin, Nicolas, 142, 156–7
Fraunhofer, Joseph, 110, 120
friction machine, *242*
Frisius, Gemma, 28

Galileo, 74–6, 109, 134, 146
Gama, Vasco da, 22
Ganger, Nicholas, 138

gasometer, 157–8
gas thermometer, 150
Gatty, Joseph, 140
George III, King, 25, 49, 71, 86, 129, 147
German silver, 174, 186
Gilbert, John, *243*
gilding, 186
globes, 100–2, *104,* 108
Godfrey, Thomas, 64
gold, 171
goldsmiths' scales, 242
goniometer, *216*
gores, 101
Gould, John, 64
gradiometer, 46
Graham, George, 43, 103
graphometer, *32,* 33–4, 47
Great Exhibition of 1851, 95, 179
Green, James, 141
Greenough, Thomas, 47
Gregory, James, 79
Griffin & Tatlock, 149
groma, 20–1
Grubb, Howard, 111
Gunter, Edward, 24
gyro-compass, 52

Hadley, John, 45, *56,* 57, 64, 81, 86
Hagger, William, 64
Hall, Chester Moor, 86
Harirot, Thomas, 75
Harrison, John, 60–1, 156, 242
Hassan, Abu'l, 67
Hauksbee, Francis, 138, 154
Hearne, 81
heliostat, 77
Helmholtz, H. von, 168

INDEX

Henry the Navigator, 50
Hero of Alexandria, 21, 23, 38
Herschel, John, 114
Herschel, William, 85–6, 93, 110, 173
high range thermometers, 149–53
Hilger brothers, 111
Hill, Nathaniel, *14*, 108
Hindley, Henry, 43
Hobson, H., 153
hodometer, 23, 239
Hoefnagel, George, 118
Holland circle, 28
177
Hooke, Robert, 80, 120–4, 137,
Hopkins, W. R., 141
Horton, T., 138
Houghton, Rowland, 46
Howe, Joseph, 78
Huggins, George, 89, 90, 93
Huggins, William, 112–13
Hulot, M., 178
Huygens, Christian, 75–7, 81
hydrometer, 65

immersion lenses, 131
industrial thermometer, 149
inhaler, 169
instrument makers, 13–19, 138–9, 176–9
inverted microscope, 131
iron, restoring, 186–7
ivory, 171–2, 187

Jacks, James, 47
Jackson, Joseph, 85
Jones, Thomas, 90, *91*, 161
Jones, W. & S., 49, *97*

Kennard, John, 48
keratometer, *219*
King, Benjamin, 64
Knight, Gowin, 52

lacquering, 185–6
Lamb, Anthony, 63
Langlois, 162
laryngoscope, 168
Lassell, William, 94
lathes, 43, 177–80
Lavoisier, 157
lead lines, 50
leather, refurbishing, 173, 192
Leeuwenhoek, A. van, 118, 120
Leibniz, G., 143
Lemaire, J. P., 98
lenses, 74–6, 79, 81, 86, 88, 92, 110, 119, 120, 128, 131
Leroy, 61
level, *39, 40,* 41
Lewis, John, 46
linear measurement, 21
Lippershey, 74
liquid compass, 52
Lister, Joseph, 128
lithotrite, 168
lodestone, 50
logarithms, 59
log boards, 50, 54
longitude, 59–60
low temperature thermometers, 150

magnetometer, *220*
Mann, James, 78, 81, 86, 122, 178
marine barometer, 140
mariners' compass, 52
Marshall, John, 78, 122–6, 128, 178

INDEX

Martin, Benjamin, *30,* 105, 107, 126, 128
Martin, Johann, 71, 238
Martin, Joshua, 88, *89*
Maudslay, Henry, 178–9
May, C., 95
Mayer, J. T., 45
Mégnié, 156
Merrill, P., 48
metal mirrors, 85, 94, 96
micrometer, 46, 128
microscopes, 13, 17, 19, 45, *85,* 116–33, *121, 123, 126,* 142, 188, 192–5, *241*
microspectroscope, 115
mining dial, *212*
mirrors, 79–81, 85, 94, 96, 126, 130–1, 193–5
Moore, Henry, 72
Morgan, Francis, *84*
Morris, M., 64
Moxon, 162
Mudge, John, 85
Mudge, Thomas, 60
Muntz metal, 173–4
museum microscope, 129
Musschenbroek, 118

Nairne, Edward, 90, 128, 140
Nasmyth, James, 93–4, 176, 179
Nauden, Hulot & Cie, 144
navigation, 13, 15, 22, 50–66, 140
needle holder, 166
Negretti & Zambra, 139–44
Newman, J., 151
Newton, Isaac, 75, 78–80, 109, 147
nickel, 174
nocturnal, 55 *56*

Norden, John, 28
Norwood, Richard, 59

octant, 15, 34, 45, *56,* 57, 244
odometer, 23
opera glasses, 99
ophthalmometer, 169
opthalmoscope, *168*
optical square, 29, 31
Ordnance Survey, 27–8, 50
orrery, *76,* 101, *102–3, 105,* 106–8

pantograph, 162, 244
Paquelin, André, 169
parallel rule, 161
parchment, 193
Paré, Ambroise, 164
Pascal, Blaise, 135
Pastorelli & Rapkin, 149, 151
patent log, 65
pedometer, 24
perambulator, 23
perspective glasses, 77
perspicillum, 75
phosphor bronze, 175
photography, 49, 113
plane tabling, 34–5, *36, 48*
planetarium, *105,* 107
platinum resistance thermometer, 152
pneumatic machine, *136*
pocket dial, 68
pocket globe, 108
pocket microscope, *120*
poke dial, 69–71, 238
Pope, Joseph, 107
portable barometer, *138*
Powell & Leyland, 131

precision balances, 154–8, *155*, 242
prismatic binoculars, 99
probes, 167
proportional compasses, 162
Pyefinch, Henry, *87*, 88
pyrometer, 152–3

quadrant, *15*, 22, *33*, 34, 43, 48, 57, 63, *98*, 238
quart de cercle, 33

Ramage, John, 93
Ramsden, Jesse, 25, 27–8, 40, 43, 89, 90, *91*, 139, 141, 156–7, 239, 240
Rathbone, Aaron, 28
Reeves, Richard, 78–9, 122
reflecting circle, *44*, 45
reflecting microscope, 130
reflecting octant, 57–9
refractometer, *229*
regulator, 149
repeating circle, 45
Riddel, 131
ring dial, *68*, *70*
Rittenhouse Brothers, 49, 93, 107
rods, 25–7
Ross, Andrew, 128, 131
Rowley, John, 103, 105
Roy, William, 25, 27, 40
Rue, Warren de la, 94
Rugendas, Nicholas, 71
Russell, John, 108
Rutherford, John, 151

salinometer, 65
saws, 164
Scarlett, Edward, 78, 81, 86, 141

Scheiner, C., 162
Schissler, 71
Schniep, Ulrich, 68
Scott's expedition, 35, 46
screw-barrel microscope, 118 *119*
screws, 178–9
sectograph, 161
sector, 36, *37*, 161
seismograph, *230*
self-registering thermometer, 150–2
Selligue, 128
Seneca, 117
Senex, John, 108
set square, 160, 161
sextant, 22, 31, 50, *54*, *57*, *58*, *62*, 64, 239, *240*
shagreen, 173, 192–3
shepherd dial, 73
Short, James, 81–5, 89
Shuttleworth, 128
Siemens, 152–3
silver, 171–2
silver gilt, 186
silver plate, 175–6
simple microscope, *117*
Sinclair, George, 139
Sisson, Jonathan, 38, 63, 142
Six, James, 150–1
Smith, Addison, 88
sounding machine, 65
spark spectroscope, *114*
spectroscopes, 109–15, *111*, *112*, *113*, 242
spectrum analysis, 110
spherical dial, 73
spherical lens, 123
spirit level, 38, 46
spring balances, 158

station pointer, 65
steel, 176
Steinheil, 120
sundials, 67–73
surgical instruments, 164–70, *165, 167*
surveying, 15–16, 20–49
surveyors' wheels, 23–4

tachometer, *233*
telemeter, *234*
telescopes, 13, 17, 19, 38, 41, 45, 58, 74–99, *78, 79, 82, 83,* 110, 173, 192
Thaxter, Samuel, 47
theodolites, *16,* 28, 33, 38, *39,* 40–9, *42, 83,* 92, 239, 240
thermometers, 18, 90, 141, 146–53
Thevenot, 38
Thomson, William, 51
Thury, C., 25
Tiedemann, 128
Timby, T. R., 141
Tompion, Thomas, 43, 103
Tonnelet, 149
Torricelli, E., 134
transit instruments, *80, 91,* 92, 94, 97
triangular compasses, 162
triangulation, 22–8, 35, 40
Troughton & Simms, 28, *42, 44,* 90, 92, 94, 111, 157, 239
T-square, 160

Tulley, Charles, 92
Tuttle, Thomas, 71, *163*

urethroscope, 170

Varley, Cornelius, 131
Vaucanson, Jacques de, 177
Veitch, James, 93
Vernier, *31,* 32, 38, 43, 45, 141–2
Vidie, Lucien, 143–4
Vinci, L. da, 162
Vitruvius Pollio, 23, 38
Voigt, Henry, 93
Voigtländer, 98
Volckmer, Tobias, 71

water level, 38
Watkins, Francis, 88, 128
waywiser, 23–4, *23,* 239
Wenham & Zeiss, 128–31
Wheatstone, Charles, 73, 131
Whitworth, Joseph, 179
Willebrand, J., 71
Wilson, Alex, 148
Wilson, James, 118
Winter, Thomas, 129
With, George, 96
Wollaston, William, 119
Woodruff, L., 141
woods, repairing and refurbishing, 172–3, 187–91
Wright, Thomas, 38, 105, *106*

Yarwell, John, 78